T0249877

guide to concrete
dike revetments

CENTRE FOR CIVIL ENGINEERING RESEARCH AND CODES
TECHNICAL ADVISORY COMMITTEE ON WATER DEFENCES

Taylor & Francis
Taylor & Francis Group

LONDON AND NEW YORK

Published by Taylor & Francis
2 Park Square, Milton Park, Abingdon, Oxon, OX14 4RN
270 Madison Ave, New York NY 10016

Transferred to Digital Printing 2007

ISBN 90 376 0004 2 (pbk)
ISBN 90 212 6062 X (hbk)

CUR
Postbus 420
2800 AK Gouda
The Netherlands

Publisher's Note
The publisher has gone to great lengths to ensure the quality of this reprint
but points out that some imperfections in the original may be apparent

PREFACE

The "Guide to concrete dike revetments" was published in 1984 (in Dutch) under the auspices of the "Technical Advisory Committee on Water Defences" (TAW) and the Centre for Civil Engineering Research and Codes (CUR). This publication is a translation of the above mentioned report and, therefore, reflects the state of the art at the time of first publication in 1984.

On the initiative of the contact-group "Wet hydraulic engineering", which includes representatives of the Dutch cement industry, the "Rijkswaterstaat" (The Netherlands Public Works Department), the Delft University of Technology and the Wageningen University of Agriculture, a decision was made to commence a study on the subject of concrete block revetments on dike slopes, in co-operation with TAW and CUR. A study group was formed for that purpose in 1981, which – in TAW connection – is set up by Working Party 4 "Dike revetments" and is indicated as Working Party B "Guide to concrete dike revetments" and – in CUR connection – operates as research committee C 45 "Concrete dike revetments". The Committee had the task of writing a guide concerning design, construction, management and maintenance of concrete block revetments. The reasons for the study included, inter alia:

- there were no specific requirements formulated at that time;
- the possibilities for application (of blocks) are determined by the product systems placed on the market by the manufacturers; only experienced organisations and management in this field can pass judgement on the applicability of these systems;
- the existing gaps in the knowledge can lead to over- or under dimensioning;
- concrete block revetments can in future become economically more attractive than asphalt revetments and, especially, natural stone (rock) and could therefore be applied more frequently.

The results of the study have been presented here in the form of a guide. This guide is primarily aimed at those technical staff directly involved in design and management, and employed by district water boards, consulting engineers, provincial public works departments and The Netherlands Public Works Department (Rijkswaterstaat). The guide is *not* intended to be a scientific publication in which theoretical fundamentals are thoroughly discussed. It has been attempted to present as much as possible of the background information without providing a solution for each possible problem. The latter is, in connection with different circumstances (geographical as well as other aspects), not only impossible but also undesirable, because it could result in an inflexibility towards other solutions.

The matters discussed in this guide need to be viewed as a summary of criteria which have to be satisfied in the design of a concrete block revetment. The accent here is more on the behaviour of a concrete block revetment under the influence of hydraulic loads

3

and on guidelines for the execution of the construction, than on the considerations of concrete as a material.

In writing the guide, use has been made of recently carried out research at the Delft Hydraulics Laboratory and the Delft Geotechnics. However, that research has not yet progressed to the extent that the strength of various types of revetment are known completely. This guide will, therefore, have to be modified in due course.

In order to keep the guide (comparatively) concise and readable, many points have been treated rather briefly. More background information can be obtained from the CUR/COW report "Background to the guide to concrete dike revetments" obtainable from COW (Centre for Flood Defence Research). In that report the overall problem has been approached more theoretically and extensive references to literature have been included. It should, however, be stressed that this guide can be used as an independent unit, without needing the assistance of the other report. Both the guide and the background report were written by Ir. G. M. WOLSINK.

The committee, at the time of publication of the guide, comprised the following members:

Prof. ir. A. GLERUM, Chairman	Delft University of Technology
Ir. G. M. WOLSINK, Secretary	Delft University of Technology
C. C. BAKKER	De Hoop B.V.
Ir. W. BANDSMA	Rijkswaterstaat, Division of Road Construction
Ir. H. BURGER	Rijkswaterstaat, Directorate Sluices & Weirs
Ing. M. C. P. COK	Gebr. Van Oord B.V.
Ir. E. H. EBBENS	Rijkswaterstaat, Centre for Flood Defence Research
Ing. T. J. LEENKNEGT	Provincial Public Works Department of Zeeland
Ing. M. G. M. PAT	Delft University of Technology
Ing. L. A. PHILIPSE	Water Board Fryslân
Ir. K. W. PILARCZYK	Rijkswaterstaat, Delta Department
Ing. J. P. VAN DER REST	Drainage District Rhine and Yssel
C. ROOK	Concrete factory "De Hoorn" B.V.
Ing. A. ROOS	Rijkswaterstaat, Directorate of North Holland
Ir. A. P. VAN VUGT	First Netherlands Cement Industry (ENCI) N.V.
Ir. P. C. MAZURE (mentor)	TAW
Ir. J. C. SLAGTER (mentor)	CUR

At the commencement of the committee work, Ing. W. DROOGER formed part of the Committee and was succeeded by Ing. M. C. P. COK in 1982. Ing. J. P. VAN DER REST became a member of the committee in the same year.

Many thanks are due to Ing. L. J. WEIJDT and Ing. W. DROOGER for their practical advice on many occasions.

December 1989   Centre for Civil Engineering Research and Codes
                                 Technical Advisory Committee on Water Defences

CONTENTS

5

LIST OF SYMBOLS

b	thickness of the filter layer
C	wave celerity
C_0	wave celerity in deep water
C_g	group velocity of waves
D	diameter of the sieve aperture for granular filters
d	diameter, water depth, thickness of concrete blocks
E	energy density
$F_S(S)$	probability distribution function of the loading
$F_R(R)$	probability distribution function of the strength
$f_s(s)$	probability density function of the loading
$f_r(r)$	probability density function of the strength
f	wave frequency
g	gravitational acceleration
H	wave height, vertical water level variation
H_s	significant wave height
I_c	consistency index
I_p	plasticity index
k	permeability of the filter layer
k'	permeability of the block revetment
L	wave length
L_0	wave length in deep water
n	ripening factor
O	largest aperture in geotextile filters
R	strength of the structure
S	load on the structure
T	wave period
$\bar{T}$	mean wave period
T_s	significant wave period
X_n	base variables
Z	reliability function
α	slope angle
γ	safety coefficient
Δ	relative density of the revetment material (blocks)
λ	dispersion length
ξ	wave-breaking parameter
ϱ_b	volumetric mass of the revetment
ϱ_w	volumetric mass of water
$\Delta\Phi$	maximum difference in hydraulic head below and above the block revetment

ABBREVIATIONS

COW   Centrum voor Onderzoek Waterkeringen
       (Centre for Flood Defence Research)
CUR   Civieltechnisch Centrum Uitvoering Research en Regelgeving
       (Centre for Civil Engineering Research and Codes)
LGM   Laboratorium voor Grondmechanica (Delft Geotechnics)
NAP   Normaal Amsterdams Peil (Standard Amsterdam Datum)
RWS   Rijkswaterstaat (The Netherlands Public Works Department)
TAW   Technische Adviescommissie voor de Waterkeringen
       (Technical Advisory Committee on Water Defences)
WL   Waterloopkundig Laboratorium (Delft Hydraulics Laboratory)

CHAPTER 1

INTRODUCTION

The revetment of an embankment will be defined in the context of this guide as that part of the total covering layer that is loaded directly by the waves. Under the revetment is an underlayer of clay, granular material or bitumenised sand which in the more modern embankments forms the protection for the sand body of the core. An intermediate layer sometimes exists between cover and underlayer.

In order to meet the requirement of sandtightness for the cohesionless granular material, geotextile cloths or sheets can be added to the construction.

The purpose of the revetment of an embankment is, in conjunction with the underlayer, to protect the body of the embankment against erosion due to waves, currents and other more particular loads such as ice drifts.

The attention in this guide will be mainly directed to sea walls (embankments along the sea); only occasional reference will be made to lake and river dikes.

The first use of concrete in applications to strengthen and protect sea defences dates back to the beginning of this century. First use was probably to fix old and worn slopes of hand-placed natural stone (or rock) by filling holes with concrete. This produced a dense and closed surface, but also made a rigid monolith which left the whole construction no freedom of movement so that any loss of subsoil material could lead to breaks. The original flexibility of the stone cover was thus lost.

The results obtained with concrete were, in general, initially not satisfactory, which could be ascribed to the rigidity of early applications but also to the inadequate knowledge of the appropriate concrete mixes to achieve a strong and dense concrete. This has caused many disappointments, especially when fresh concrete, placed in situ, came into contact with seawater (below and immediately above high water). The slopes thus treated began to look shabby after a short period of time.

Precast concrete blocks were in the past also often manufactured on site. Again as a result of insufficient knowledge of concrete as a material and careless production (unskilled labour, inadequate equipment, etc.), the blocks were repeatedly corroded and damaged by wave impact. Finally, such blocks would fall apart and return to loose aggregate state.

In choosing the shape of blocks, a lot of care was paid in the past to a dense and yet flexible link between the elements of the revetment, in order to give the slope construction sufficient elasticity to permit some settlement of the foundation. Systems which did exist in the past as marketable units with all sorts of refined shapes to interlock have largely disappeared for economic reasons. The complicated shapes made mechanical manufacture often difficult if not impossible, whilst during the revetment construction the placing of such blocks was equally difficult.

From the start of concrete block application until about 1965, the placing of blocks remained almost exclusively a hand-operation. Each block was placed separately, by one or two men depending on the weight, by hand on the slope. The rate of production of such revetments, using the larger blocks, was not great. As the construction works increased in size, made possible by faster delivery of sand (for bank cores) and greater mechanisation to reduce labour costs, a search had to be made for a faster placing

method. It is the use of special blockclamps, with the ability to handle several blocks in one operation, which permitted greater production per working day by placing several blocks together on previously prepared parts of the bank slope.

Therefore the manufacture of concrete slope revetments or elements, in situ where they are needed for construction, and especially pouring concrete in the tidal zone, has to be discouraged according to experience. Only when production takes place in concrete block factories where the mixed concrete can be pressed, vibrated or shaken, satisfactory results can be expected from normal concrete. Casting concrete in situ in the tidal zone can exceptionally be tolerated as a result of more recent developments in the field of special "under-water concrete".

In this type of concrete, special additives and/or preparation produce a greater resistance against segregation during pouring (e.g. colloidal concrete).

The choice (type and dimensions) of the concrete block revetments completed so far depends only on experience factors and on personal judgement or preference. Objective design criteria have not been available. This means that in situations where little or no experience has been gained, i.e. for extreme loading conditions (super storms), the question can be put whether the design is in fact technically and economically the correct one (neither too light nor too extremely heavy).

It is the purpose of this guide to contribute to a more reliable method of design for a concrete block revetment. However, due to the complexities of the subject matter, there are as yet no simple computer models available to calculate the revetment stability under wave attack. Research for this purpose at the Delft Hydraulics Laboratory and the Delft Geotechnics will perhaps in the long term produce a practically useful approach. In the meantime it is possible to use with advantage the results of tests carried out in a wave flume as presented in Chapter 13. For the larger bank sections, it may be useful to test the concerned section with revetment, together with the corresponding boundary conditions, on the largest possible scale in a wave flume.

In case of a new type of revetment, it is furthermore recommended to test the revetment by means of test sections under practical conditions, if possible. The loading conditions under which the revetment has failed can then be determined, provided that an underlying hidden protection was applied.

CHAPTER 2

REVETMENT REQUIREMENTS TO BE MET

2.1 Functional requirements

In accordance with the purpose of the revetment on a sea wall, i.e. to form a protection for the body of the embankment, the following functional requirements need to be stated:

a. The revetment must be able to resist:
   - the combination of wave and flow attacks;
   - the forces exerted by driftice, ice frozen to the revetment, etc.;
   - excess pore water pressures due to a raised phreatic (groundwater) level.
b. The underlying soil particles must be retained; the revetment, together with any filter construction present, must prevent migration of these soil particles. The revetment must also protect the filter construction against wash-out.
c. The revetment has to be durable, i.e. it has to resist erosion due to materials being carried in the flows over it (sand, gravel, pieces of natural stone, etc.) and against frost and chemical action.
d. In order to fully and permanently serve its purpose, it is important that the revetment can mould itself to possible form changes of the slope (settlement and/or scour) without destroying the bond of the revetment surface. If the subsoil settles locally, or is eroded, or disturbed by animal burrows, and the revetment cannot adjust to the new contours, hollows are created under the revetment so that the construction is seriously weakened, for instance under an external loading of breaking wave impact. A strong interlock between concrete elements, preventing settlement of the elements into hollows underneath, can be a disadvantage in this context (see also section 7.1).
   On the other hand, such strong interlock between elements on a sound foundation will provide greater stability under wave attack than non-interlocking blocks (elements which derive stability from mutual friction) can provide.
   Flexibility and stability requirements under wave attack are thus seen to be contradictory. In practice the best possible compromise solution would have to be adopted.
e. The whole of the revetment and subsoil must be stable against sliding.

2.2 Requirements for technical execution

In order to achieve the optimum construction costs, the following requirements could be framed:

14

a. The revetment has to be quick and easy to place, preferably by mechanical means. For the construction of sea defence works, a limited period of the year (April to October) is available (outside the storm season). For the zone of daily tides the requirement for rapid placement is even more obvious.

b. The revetment has to be such that setting out and measurement can be carried out easily, especially for non-straight dike alignments.

There are, however, systems of revetment with units shaped for very accurate placing, because the tolerances are virtually nil. A typical example is illustrated in Fig. 1. This system does not allow deviations from a straight line. Curved work in such cases can only be executed with in-situ connections by pouring fresh concrete between short lengths of straight alignments. This does not look particularly attractive, whilst the concrete quality as well as the behaviour under loads and settlements can be different.

By leaving joints "open" it is possible with many systems, within certain limits, to place units around a longitudinal curve. If the curves are reasonably easy, such open joints in a revetment on a granular filter of sufficient coarseness need not give rise to problems. A clay underlayer is less suitable in these situations.

Any camber, i.e. a curved slope in the vertical sense, needs to be small so that the settlement of blocks is not obstructed, as that could reduce the interlock and consequently the stability of the blocks.

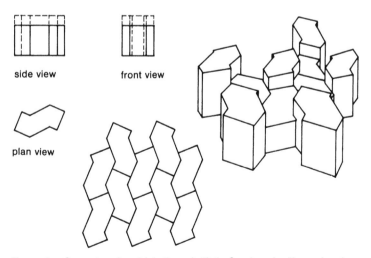

side view   front view

plan view

Fig. 1.   Example of a system in which there is little freedom in dimensional co-ordination.

Mechanized placing of concrete units.

2.3 Management and maintenance requirements

For a revetment to meet the requirement to provide a durable protection, the following demands need to be satisfied:

a. If unexpected damage occurs locally, it is important that the revetment can be repaired quickly and easily.
b. The revetment must not be liable to damage by vandals.

2.4 Special requirements

Local circumstances could lead to one or more of the following requirements.

a. Where an embankment is subjected to frequently occurring wave attack, it can be useful to reduce the wave run-up on the top of the revetment by means of specially shaped blocks (with projections or voids).
b. Temporary revetments should as much as possible be of block types which can be reused elsewhere.
c. Special requirements on the surface treatment of the revetment in order to fit better into areas of environmental importance.
d. The revetment together with the foundation layer sometimes has to be watertight when embankments have to stand up to high water levels for longer periods of time.

CHAPTER 3

REVETMENT TYPES

3.1 General

For the choice from various alternatives in a given situation, the assessment criteria (technical and financial) to be formulated should comply with the requirements outlined in Chapter 2. Because the various technical criteria are not all of equal importance in regard to the final choice, their significance will have to be weighted. Finally, the various alternatives will be checked against the assessment criteria and the weighting factors, thus producing a technical valuation. For further considerations on this method, the reader is referred to the CUR/COW report "Background to the guide to concrete dike revetments" [32].

For the classification of concrete block revetments, various aspects can be selected, such as the following main division:

- according to block shape;
- according to the degree of permeability;
- according to the relationship between concrete block revetments and the permeability of the underlayer and intermediate layer;
- according to the combination of prefabricated concrete mats with artificial fibre carriers or cables.

The following could be mentioned as subdivisions:

- the possibility of mechanical placing;
- reinforced or mass concrete;
- in-situ concrete or precast concrete;
- above water or both above and below water construction.

3.2 Main division

3.2.1 *Shape of concrete elements*

The elements can be subdivided into:

- block and column shaped elements;
- slab shaped elements;
- uninterrupted plate shape (continuous monolithic pour).

Block and column shaped elements
Mainly these shapes are used in the Netherlands. The block shaped elements can be subdivided in accordance with the given shape, the interlock between blocks in relation

18

to the possibility of comparative movement of one block to another, the locking action, the degree of wave run-up reduction provided, etc.

Block shaped elements are usually made with gravel aggregate, but also with basalt aggregate to increase the unit weight. For aesthetic reasons the blocks can be washed when freshly cast to show exposed aggregate, or sometimes provided with an additional basalt-aggregate surface layer for different aggregate exposure. The most common block element dimensions in plan are 0.50 m × 0.50 m or 0.25 m × 0.30 m, but 0.30 m × 0.30 m and 0.25 m × 0.25 m also exist. Matching thicknesses usually range between 0.15 m and 0.30 m. Blocks are commonly placed on the slope in a staggered vertical-joint pattern (see top left-hand photograph page 20).

The column or polygon shaped elements can, depending on the producer, be regular or irregular in shape. These elements derive their strength as revetment material mainly from the joint filling material which has to take care of the interlock friction. The columns should therefore be made such that joint fillers can be applied.

Various revetment types with interlock, obtained by elements "hooking" together, were put on the market in the past. Due to problems with dimensioning accuracy and automatic fabrication, this type is no longer in use*.

The height of column-shaped revetment blocks ranges in practice from 0.20 m to 0.50 m, and is also expressed in weight per m². Joint filler materials used are gravel, broken stone, waste and non-hydraulic slag. Maximum dimensions of the individual joint filler granules should be in accordance with the average spaces between columns. A spread in grain sizes of the joint fillers produces better results against waves washing the joints clean.

Slab shaped elements

Solid slab shaped elements are rarely if ever used in the Netherlands. However, slab shaped blocks with holes for vegetative growth are used on a large scale. In addition to the holes there are also shallow grooves in the upper surfaces, linking the holes. If all holes and grooves are filled with clay** and all clay is thoroughly rooted with grass, a strong revetment could result. The slabs are usually 0.40 m × 0.40 m or 0.40 m × 0.60 m in plan dimensions with thickness ranging from 0.09 up to and including 0.15 m. This type of revetment can be placed by hand as well as mechanically, and is commonly applied well above the influence of the daily tide.

Continuous concrete slab

The continuous concrete slab is rarely if ever used in the Netherlands because of the danger of uneven settlement and soil erosion. It is still used on a large scale abroad.

---

* In the Netherlands
** Clay is used in the Dutch text and it may be Dutch practice, although it may not be the best soil for growing grass.

Some types of revetment.

20

3.2.2 Degree of permeability of concrete revetments

The revetment can be:

- closed or nearly closed;
- open.

The permeability of tightly fitting blocks can be regarded as very small in comparison with the permeability of the underlayer. To make such revetments more pervious, the block shaped elements can be supplied with chamfered corners and/or notches in the sides of the blocks. That will moreover permit filling with joint material to provide additional interlocking action.

Permeability of column shaped elements with joint filler materials (granular) is large in comparison with the underlayer. Hence the revetment is classified as very open. If that type of revetment is penetrated with a hot poured bitumen, a completely closed revetment will result.

3.2.3 Relationship between concrete block revetments and the permeability of the underlayer and intermediate layer

How the hydraulic loadings act on the revetment cannot be seen in isolation from the mutual relationship of the permeabilities of the underlayer, the intermediate layer and the revetment (see also Chapter 13). Accordingly, the selection of the concrete block revetment is largely determined by that relationship.

The following differences can be noted (see Figs. 2 to 9):

- open revetment on permeable intermediate layer and impermeable underlayer;
- open revetment on permeable intermediate layer and permeable underlayer;
- open revetment without intermediate layer on permeable underlayer;
- continuous revetment with or without permeable intermediate layer on permeable underlayer;
- continuous revetment on permeable intermediate layer and impermeable underlayer;
- continuous revetment without intermediate layer on impermeable underlayer;
- closed revetment on impermeable intermediate layer and impermeable underlayer;
- grassed revetment on impermeable underlayer.

Figures 2 to 9 offer only some examples of possible types of construction, without attempting to be exhaustive; they should therefore not be interpreted as standard constructions.

Because blocks laid directly on an impermeable underlayer (clay) are more stable under wave attack, this type of construction may be preferred. It is, however, not always possible or justifiable to apply that construction.

At the lower levels in the tidal zone the erosion resistance of clay and the construction operation for a placed type of revetment are problematic, whilst in addition, in cases of

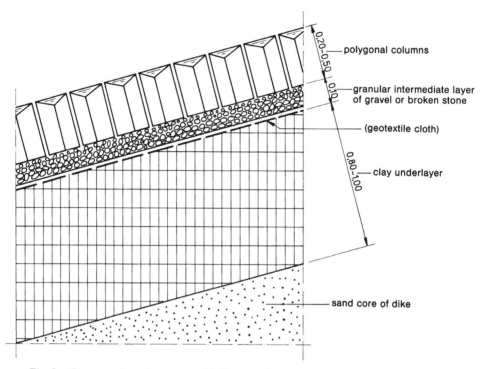

Fig. 2. Open revetment on permeable intermediate layer and impermeable underlayer.

In Fig. 2 the following labels appear:
- polygonal columns (0.20–0.50)
- granular intermediate layer of gravel or broken stone (0.10)
- (geotextile cloth)
- clay underlayer (0.80–1.00)
- sand core of dike

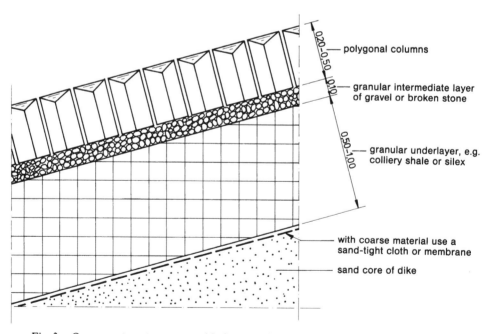

Fig. 3. Open revetment on permeable intermediate layer and permeable underlayer.

In Fig. 3 the following labels appear:
- polygonal columns (0.20–0.50)
- granular intermediate layer of gravel or broken stone (0.10)
- granular underlayer, e.g. colliery shale or silex (0.50–1.00)
- with coarse material use a sand-tight cloth or membrane
- sand core of dike

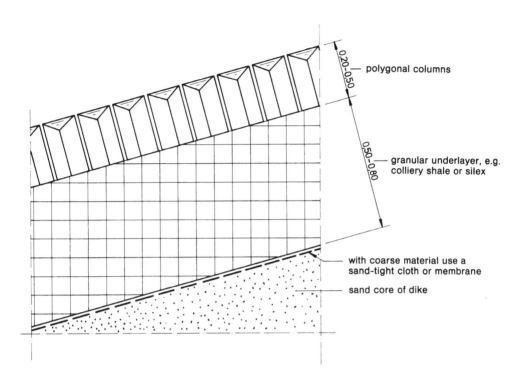

polygonal columns

granular underlayer, e.g. colliery shale or silex

with coarse material use a sand-tight cloth or membrane

sand core of dike

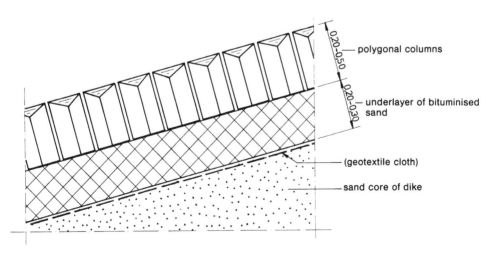

polygonal columns

underlayer of bituminised sand

(geotextile cloth)

sand core of dike

Fig. 4.   Open revetment without intermediate layer on permeable underlayer.

23

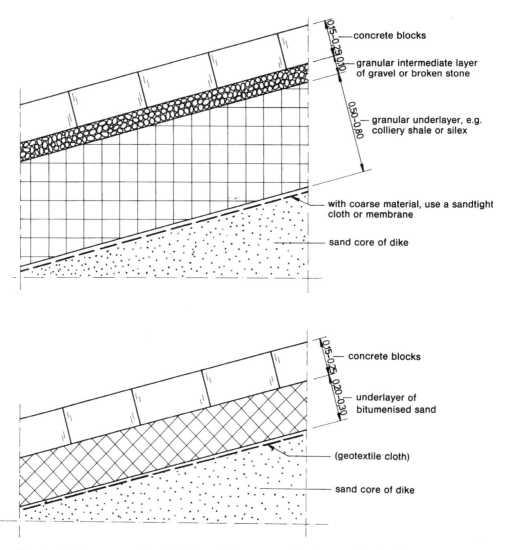

concrete blocks

granular intermediate layer
of gravel or broken stone

granular underlayer, e.g.
colliery shale or silex

with coarse material, use a sandtight
cloth or membrane

sand core of dike

concrete blocks

underlayer of
bitumenised sand

(geotextile cloth)

sand core of dike

Fig. 5. Continuous revetment with or without permeable intermediate layer on permeable
underlayer.

possible porewater pressures in the body of the dike, a permeable underlayer is essential. The necessary requirements for clay are described in sections 5.2 and 5.4.1.
Where in the Figs. 2 to 9 reference is made to geotextile *in brackets,* it may also be omitted from the construction. In connection with site construction work the geotextile can provide a cover to make placement of the subsequent layer easier; the geotextile performs in that case a protection role – a separator against accidental mixing of materials.

24

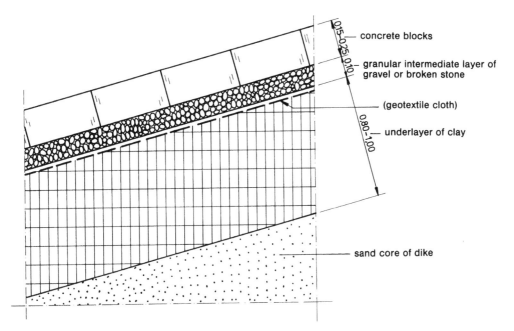

Fig. 6. Continuous revetment on permeable intermediate layer and impermeable underlayer.

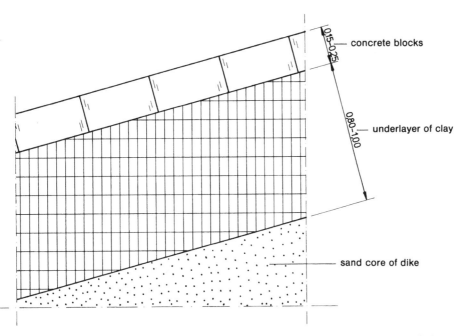

Fig. 7. Continuous revetment without intermediate layer on impermeable underlayer.

25

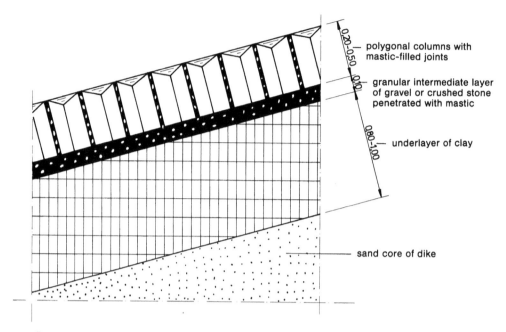

0.20-0.50 — polygonal columns with mastic-filled joints

0.10 — granular intermediate layer of gravel or crushed stone penetrated with mastic

0.80-1.00 — underlayer of clay

sand core of dike

Fig. 8. Closed revetment on impermeable intermediate layer and impermeable underlayer.

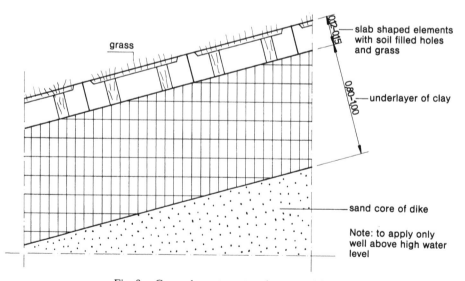

grass

0.12-0.15 — slab shaped elements with soil filled holes and grass

0.80-1.00 — underlayer of clay

sand core of dike

Note: to apply only well above high water level

Fig. 9. Grassed revetment on impermeable underlayer.

26

3.2.4 *Prefabricated mats of concrete blocks on geotextile carriers*

A new development in the protection of river and lake banks comprises the application of factory-made interlocking concrete blocks on a carrier of geotextile fabric and sometimes provided with cables through the blocks.

Application so far is mainly on the banks of canals, rivers and lakes. These mats are permeable and, therefore, have to be considered as open revetments.

3.3 **Subdivision**

3.3.1 *The possibility of mechanical placing*

There is increasing preference for the use of systems which permit mechanical construction methods. One result of this is the development of the prefabricated block mats described in section 3.2.4 above.

Originally only the block shaped elements could be placed mechanically. At present column shaped elements can also be placed mechanically, so that it can be stated that in the Netherlands most of the currently used revetments can be placed mechanically.

3.3.2 *Reinforced or mass concrete*

Both block shaped and column shaped revetments, as well as slab shaped elements with holes for vegetative growth are manufactured in mass concrete.

It is only for special revetment constructions that reinforced concrete is used in the Netherlands.

3.3.3 *In-situ concrete or precast concrete*

Precast concrete blocks should have preference over in-situ placed concrete because the former ensures in general a better quality concrete. Even made-to-measure gap-fill blocks should preferably be precast. In general the use of in-situ concrete even for filling joints and narrow gaps should be avoided as much as possible.

3.3.4 *Above water or both above and below water construction*

Only the precast block mats can be placed under water; in the absence of a lower boundary however. All other block or column revetments are limited to above water application.

Another new development consists of colloidal concrete for pouring under water.

CHAPTER 4

CONCRETE TECHNOLOGY

4.1 General

During its life in use as dike revetment, concrete quality can deteriorate. The causes for this can be mechanical, biological, physical or chemical.

Mechanical attack on concrete can be due to excessive loading or in the case of revetments mainly through the scouring action of sand and water. Biological effects arise in the form of growth of algae, (water) plants and other organisms. These growths may adversely affect the possibility of walking over the revetment, and the aesthetics, but do not (in general) damage the concrete skin. Physical causes of damage are for example large temperature changes and frost. Chemical attack can occur through any contact with surface water, seawater, rainwater, or pure water.

Corrosion of steel, if present in the concrete, can be destructive and should be prevented, just as any other form of attack should be guarded against.

In the choice of the raw materials for the construction elements, important considerations are the workability of the concrete mix and the economic aspects.

4.2 Characteristics of wet concrete mixes and concrete

The degree of compaction of a concrete mix depends on many factors, such as type and quantity of cement, particle shape and distribution of the aggregates and especially the amount of water with possible additives in the mix.

Water-separation or "bleeding" of a concrete mix can be counteracted by using less water, including additives in the mix and by increasing the "specific surface" of the aggregates (i.e. total surface of all particles).

Strength of a concrete element of a construction is most effectively tested by a destructive test. It is possible to use laboratory tests instead. Standard laboratory tests exist for the determination of erosion resistance, modulus of elasticity, shrinkage, creep and other characteristics of concrete. By choosing heavy aggregates the volume mass of concrete can be increased, which has a favourable influence on the stability under wave attack (see Chapter 13).

4.3 Requirements and regulations

The manufacture and application of concrete is tied to a number of regulations (see also References). Unless otherwise specified, the concrete as well as the products made of it, have to comply with the requirements prescribed in the regulations.

Poor and good concrete side by side.

Severe erosion due to the scouring action of granular material and water.

Concrete used for dike revetments has to be at least of a quality defined as B 30 in the (Dutch) regulations for concrete*, in order to limit the damage potential of mechanical, physical or chemical attacks. If strength during transport is important, greater strengths may be required.

For better resistance against chemical attack, the cement has to have greater than the normal sulphate resistance and a low content of free lime. The type of cement which meets these requirements is blast-furnace cement, which therefore is preferable.

The used concrete aggregates have to comply with the pertinent product standards. If reinforcement is used in the concrete, the cover required under the regulations for concrete for an aggressive environment (sea wall revetment!) should be strictly adhered to. Careful compaction and curing is of importance to achieve a dense water-tight surface. The concrete should be protected for a period of at least one week against drying out. Control of the composition of the concrete, pouring, curing and strength is required.

4.4 Amplification

The foregoing requirements and codes have been formulated in the light of experience with the use of concrete in water construction works. Exaggerated requirements for the strength of concrete materials can result in unnecessarily high costs.

Building materials in an aggressive environment are under attack, sometimes slowly, occasionally quickly. The rate of progress of such attack depends on the material's resistance. In that connection the composition and pore distribution is of course important. Demands are therefore made on the concrete's raw materials, fabrication, storage and quality. Depending on the type of cement, the concrete can suffer more or less from the chemical attack of calcium and magnesium sulphates present in seawater. A fortunate circumstance is that sulphate attack in seawater is much less than that shown by comparative tests in a laboratory with the same sulphate concentration. This is due to the presence of chlorides in seawater. Practice has shown that attack on concrete in seawater is mostly small, provided the concrete is dense and of good quality. The Dutch regulations for concrete in coastal areas nearly always insist on the use of blast-furnace cement. If the blast-furnace slag content is at least 65 %, this cement is sulphate-resistant.

Frost forms another type of attack. In CUR report 64 "Frost resistance of concrete" [10], test results are given which show that blast-furnace cement provides better frost resistance than does Portland cement (when subjected to a special freeze test).

Finally, there is the possibility of weathering of concrete due to various mechanisms. Mention can be made of the mechanical abrasion due to a streaming flow carrying sand and/or gravel, and swelling action when the salts in the water form crystals within the concrete.

Temperature differences can cause tensions in the surface layers, causing cracking.

---

* Concrete quality B 30 = 30 N/mm² after 28 days.

The above discussion clearly shows that concrete constructions have a limited life, but that the life can be strongly influenced by the concrete composition and the manner of mixing and curing. Standards provide clear guidance on the use of reinforced concrete. Corrosion to steel reinforcement due to penetration of carbonic acid from the air and chloride from seawater can be reduced with dense, good-quality concrete.

In addition, the concrete cover has to be adequate and crack formation should be prevented as much as possible. It should, however, be noted that steel reinforcement in concrete for dike revetments is rarely applied.

CHAPTER 5

REQUIREMENTS TO BE MET BY UNDERLAYERS

5.1 General

Each underlayer or foundation together with a possible intermediate layer can be defined as the transition construction between the core (usually made of sand) of the embankment to the outside hard revetment cover. An underlayer with an intermediate layer can fulfill the following functions, depending on the circumstances:

- prevention of soil migration (washing out) from the bank core;
- forming a smooth slope face for the easy placement of the revetment (particularly applicable to the intermediate layer);
- to form an extra safety margin in case of revetment damage, so that any consequent erosion hole does not immediately reach the sand core of the dike;
- forming a watertight layer on the permeable dike core;
- forming a good drainage layer immediately under the revetment (this applies to the intermediate layer);
- forming a temporary protective cover against flow and wave attack during construction of the dike;
- to function as a temporary slurry wall and to be retained afterwards to form a permanent part of the dike.

Underlayers can be divided into two main groups, i.e. permeable and watertight.

5.2 Watertight underlayers

A watertight underlayer is generally produced by using clay. If, however, an intermediate layer of granular material is present, possibly with a geotextile cloth or membrane to prevent the granular material being trodden into the clay, then the underlayer in relation to the revetment is a permeable layer in view of the revetment stability under wave attack.
Concrete blocks laid directly onto the clay in the zone above the tidal levels need not present any problems, provided the clay is of good quality and the revetment is a reasonably tightly fitting design. However, the use of clay in the daily tidal zone can present problems with construction as well as lead to erosion.
When the blocks fit snugly to the clay slope, it would be difficult to create hydraulic uplift pressures under the blocks. Therefore, the stability of such blocks under wave attack will be greater than for the same blocks on a granular (filter) layer. This does require an optimum fit of the blocks on the clay. Therefore the following points have to be considered:

- erosion of the clay under the revetment;
- hollows in the clay under the blocks;
- plastic and viscous characteristics of clay;
- shrinkage and swelling behaviour;
- clay workability, particularly with reference to compaction under construction conditions.

The detailed requirements for the use of clay are dealt with in section 5.4.1. This section is limited to a number of indications (derived from [20]) for the handling of clay. It should be realised that (constantly) changing work conditions make it difficult in practice to attain the optimum construction condition of the clay. In order to produce a good underlayer for the revetment, it is recommended that an attempt should be made to keep to the working limits set out below, if at all possible.
The most important aspects for the placing of clay are:

- bringing on site;
- compacting;
- producing a smooth slope and the placing of blocks.

Bringing clay on site

The optimum workability conditions, expressed in the moisture content, lie generally between comparatively narrow limits. With some types of clay these limits can, therefore, be exceeded swiftly when the moisture content changes as a result of weather changes. In this context, it is important that the plasticity index* I_p of the clay is not too low in order to eliminate the possibility of a swift change from the semi-solid state to the semi-liquid state during a period of rain ($I_p > 20$ %). A good criterion for the workability of clay is the consistency index* I_c ($I_c > 0.8$). This requirement provides the necessary difference between the plastic and the liquid limit.

Compaction of clay

During construction, efforts must be made to obtain the most dense possible soil structure (compaction). In order to obtain optimum density on site, the ideal limitations for moisture content are:

- at least equal or only a little below the optimum moisture content according to the Proctor test (see Dutch specifications, "Eisen 1978", Rijkswaterstaat);
- at maximum equal to or only a little higher than the plastic limit*.

The density attained should be at least equal to about 95 % Proctor density. Compaction has to produce a clay layer as homogeneous as possible. Layer thickness for compaction should not be too great, if necessary it can be done in 2 layers of maximum 0.40 m thickness. During and after compaction, it is necessary to carry out tests on density and

---

* See page 34.

homogeneity. Large lumps of clay that would result in non-compliance with the requirement of homogeneous compaction should not be present.

Smoothing the slope and the placing of blocks

Care should be taken during surface smoothing of the slope that the top 20 mm or so are not loosened too much, as that can lead, under the heavy loads to be expected after completion, to erosion or damage. It is equally unacceptable to fill deep vehicle tracks and suchlike with loose, crumbly clay. Vehicular or even pedestrian traffic on the finished clay slope must be avoided.

Furthermore, it is important to prevent local contamination with porous materials, because contaminations could produce local spots where uplift pressures would occur so that blocks could be lifted or erosion take place.

It is recommended that the clay layer should be placed in such a way that after compaction the whole surface is a couple of centimetres proud of the design profile. The surplus clay can then be trimmed off to produce a flat, smooth and dense surface, which should not show any cracks.

In order to improve the contact between the blocks and the clay surface, it is recommended to press the blocks down, for instance with a roller.

Work can continue when it rains until the clay is weakened to the point where placing and compaction requirements can no longer be met.

---

* Consistency limits are expressed in the moisture content (i.e. the mass of water divided by the mass of dry matter $\times$ 100 %), and split into the following:
- liquid limit W_l:
  the moisture content at which a groove made in a soil sample is just closed again after the dish containing the sample has been dropped 25 times from a height of 10 mm onto a solid surface;
- plastic limit W_p:
  the moisture content at which it is just no longer possible to roll out a ball of clay to a 3 mm thick thread without crumbling;
- shrinkage limit W_s:
  the moisture content at which the sample when drying out no longer reduces in volume, or the moisture content after which just enough water has been added to a dry piece of soil to fill all pores.

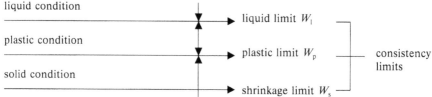

The derived indexes are:

plasticity index: $I_p = W_l - W_p$

consistency index $I_c = \dfrac{W_l - W}{I_p} = \dfrac{W_l - W}{W_l - W_p}$

in which W = water content of the soil being worked.

34

As the clay surface should not be allowed to dry out, the concrete blocks should in general be placed on the same day that the slope is trimmed to the final profile.

No blocks should be placed during frosty weather, nor when the clay is still frozen. Where a sand core embankment is constructed, with a clay underlayer for revetment foundation, the clay underlayer should, as with all other waterproof covers (e.g. asphalt-type revetment), not be too thin so that excess pore pressures in the sand core cannot lift the layers. In this context, the landward slope of the embankment should not be forgotten.

Another reason to maintain a sufficiently thick clay layer is the function described in Section 5.1, i.e. to form a safety zone when the revetment is damaged. In general (in the Netherlands), the clay thickness for sea walls is taken at 0.8 to 1.0 m.

Use is occasionally made of a paving of gravel or sand on the clay in the tidal zone in order to protect the clay during the construction stage and to make block placing easier. The method does reduce the advantage of the more stable block position on clay instead of on a granular filter. Hence this method is not recommended.

If a high-level foreland makes it necessary to extend the clay underlayer deep down, a better method would consist in creating a temporary small bank to keep the lower slope of the embankment dry. If, however, the mud and ground levels in the foreland are low, the underlayer in the tidal zone will have to be made of erosion-resistant material, e.g. colliery shale. In this situation the clay underlayer can only be provided above the daily tidal zone.

5.3 Permeable underlayers

One of the functions of a permeable underlayer is to provide protection of the embankment core against wash-out and it, therefore, has to satisfy the usual filter design criterion against migration of sand (see Section 5.5).

Permeable underlayers can be divided into:

– underlayers of granular materials, possibly combined with a geotextile cloth or membrane, or with a layer of clay underneath;
– underlayers of bonded granular materials.

5.3.1 *Underlayers of granular materials*

In order to satisfy the sandtightness requirements, the granular filter has to be constructed correctly. This filter can comprise one layer with a nearly homogeneous particle distribution, or comprise more layers with the upper layers having increasing particle sizes. The bottom layer is designed to stop particle migration from the foundation layer, and each succeeding layer has to do the same for the particles in the layer below. For a graded filter the construction will have to be a sound compromise between technical requirements and economic possibilities. It is in practice often easier and cheaper to use a geotextile cloth or membrane to achieve a sandtight underlayer.

The revetment itself also has to satisfy the filter principle concerning material tightness. The dimensions of the particles on the upper side of the underlying filter must be large enough to prevent erosion of the material through the openings (joint and holes) in the revetment. However, the application of too coarse particles of the granular material under the blocks is also wrong, because it then becomes more difficult to obtain a smooth block slope and the stoneworkers will then try to use large pieces of material to get the slope looking right. This leads to the risk of uneven settlements and an irregular surface.

Other demands to be satisfied by filters under a revetment concern the permeability. Hydraulic uplift pressures are in themselves acceptable as long as they are compensated by extra weight, so that neither the filter nor the revetment can be lifted, nor that those pressures can lead to subsoil weakening resulting in reduced shear strength and possible soil slips.

As stated before, a revetment constructed of separate unconnected elements laid directly on a less porous layer is more stable (see Chapter 13).

Various materials can be used for filter construction, e.g. various types of slag, crushed gravel, broken stone, silex, colliery shale, and waste. If slag is used, care should be taken to ensure that it does not contain water-soluble pollutants, which could be environmentally harmful.

The other function (see Section 5.1) of an underlayer to act as an in-built extra safety margin against damage to the revetment can be realised by:

- placing a clay layer under the granular layer;
- make the underlayer sufficiently thick;
- use material with a high volumetric mass;
- make the stone dimensions in the underlayer sufficiently large.

In case the base layer for any part of a revetment consists of colliery shale from a temporary bank, it is desirable to cover the shale with an intermediate layer of finer granular material, e.g. crushed stone or crushed gravel to a thickness of 0.05 to 0.10 m, for two reasons:

- if blocks are laid directly on shale, the finer particles will wash out so that voids and/ or settlements can occur;
- the unequal sizes of the shale elements make it difficult to obtain a smooth slope.

Placing a block revetment directly on a geotextile filter which in turn lies directly on the sand core of the dike cannot be recommended. The geotextile forms no suitable second line of defence in case of damage to the revetment. Also the geotextile is easily damaged during the placing of blocks.

It is a possibility to combine a geotextile filter with a granular layer in one construction. This has the advantage that no difficult filter requirements need to be met by the granular material.

The functional requirement for stability against sliding, mentioned in Chapter 2, needs to be observed by taking into account the particle shape of the granules. In order to obtain sufficient stability (especially during construction) the particles need to be angular, depending also on the degree of the slope. In general, this requirement will be met if broken materials are employed. Hence if gravel is used it should be crushed gravel.

5.3.2   *Underlayers of bonded granular materials*

For a bonded filter under the revetment, both cement and bitumen can be used as the bonding agent. With a correct selection of the mixture, both the cement and bitumen can be used to prepare an underlayer with a permeability which approaches that of cohesionless sand.

Cement-bonded underlayers have the disadvantage that they cannot follow uneven settlements and erosion holes without cracking. Bitumen-bonded underlayers are less susceptible to differential settlements due to the viscous characteristics of bitumen. On the condition, however, that settlements develop gradually. The bitumen content will nevertheless largely determine the extent of plasticity.

When a bonded underlayer is used, the revetment can be placed directly onto it. This type of underlayer also makes for a less stable block revetment than that provided by a good clay base.

5.4   **Characteristics of materials**

5.4.1   *Clay*

For the use of clay as a layer directly under a concrete block revetment, the erosion phenomenon is important in the context of block stability under wave attack. The following parameters influence the erosion behaviour of clay:

- particle dimensions;
- clay, silt and sand content;
- the clay mineral;
- the organic matter content;
- the consistency limits and the derived indices;
- manner of sedimentation and consolidation;
- the density, the (optimum) water content and the degree of compaction;
- the permeability of the clay;
- shrinkage and swelling behaviour;
- cohesion and shear resistance;
- physical-chemical characteristics of clay, porewater and eroding water;
- homogeneity.

Although it is possible to give indications as to the degree of erosion susceptibility as a function of a number of these parameters separately, there is no possibility of expressing them in an erosion formula. The description of the influence of the various parameters can in general only be qualitative.

In order to determine resistance against erosion experimentally, various types of erosion tests are possible. A distinction can be made between those tests which determine the erosion resistance along standardised procedures, taking no direct account of the circumstances in practice, and those where the relationship with the practical situation does exist. For the choice of a suitable test method, it is recommended that an experienced soil mechanics laboratory is consulted.

Although the erosion resistance of a specific clay cannot be easily calculated, certain general limitations can be indicated (see also [20]):

- minimum about 20 % clay ($d < 0.002$ mm);
- maximum about 40 to 50 % clay;
- maximum about 25 % sand ($d > 0.063$ mm);
- maximum about 3 % organic matter.

When clay is taken from saltings the degree of "ripeness" has to be taken into account. When a water-saturated soil dries out, it begins to lose water through evaporation; at first directly, afterwards through shrinkage cracks and vegetation. This drying-out process is termed "ripening". Through reduction in porewater pressures the effective stresses increase, thus making the clay more easy to work. If any saltings clay is used, it should be sufficiently ripe. This ripening process can be speeded up by placing it on a stock-pile and turning the pile several times.

Apart from the water content, the percentage clay fraction and organic matter influence the rate at which the clay "ripens" with time. The concept of the "ripening factor"* (n-number) is used. The ripening factor is defined as the water content in grams which is bound by 1 gram of the clay fraction. A completely ripened clay has a "ripening factor" n of less than 0.5.

5.4.2 *Colliery shale*

Colliery shale is a waste product from coal mining (see also [19]). During the formation of coal under high pressure, sand and clay formed sandstone and clay shale. Colliery

---

* The ripening factor can be calculated from:

$$n = \frac{W - p\,(100 - L - H)}{L + bH}$$

where

b = ratio of water binding capacity of a specific mass of organic matter to that of the same mass of clay fraction (b = approx. 3.0)

p = grams of moisture bound by 1 g non-colloidal matter (dry clay minus clay fraction and organic matter) (p = approx. 0.3)

L = grams of clay fraction per 100 g dry matter

H = grams of organic matter per 100 g dry matter

W = grams of water per 100 g dry matter

38

shale mainly consists of clay shale (this obviously applies to the Netherlands). Depending on the degree of consolidation the clay shale can be divided into clay stone and (harder) slate stone. Most of the clay shale, however, consists of clay stone, whereas the transition into slate stone is not a very distinct one. Clay stone is weak and when exposed to the air generally disintegrates easily into smaller fragments. This process can seriously reduce the permeability with lapse of time. The rate and degree of such reduction depends on the degree of consolidation. The material does not disintegrate into particles smaller than the 2 mm sieve dimension. The reason for this is that the high pressure which changed clay to clay stone removed the thin water-coating from the clay particles, and with that also removed the clay characteristics, such as plasticity, shrinkage and swelling potential. Damp colliery shale has no resistance against alternate frost and thaw cycles.

During the determination of the particle distribution, account has to be taken of a certain amount of disintegration of the colliery shale as a result of transport, storage and placing.

Compaction possibilities of colliery shale are heavily dependent on the moisture content and the grain size. When the moisture content of unsorted colliery shale from slagheaps increases by a few percent through precipitation, there is a real possibility that the optimum moisture content will be exceeded, which may lead to a state at which it cannot be compacted nor used for traffic. Water-saturated colliery shale is in principle more susceptible to settlement flows than sand. In the USA some colliery shale dams have collapsed as a result of this process.

When during the working with colliery shale the material is not too badly remoulded, the permeability will be comparable to the values for a very coarse sand (without silt), even after fragmentation of the shale. Through compaction and allowing traffic during wet weather, the resultant remoulding can cause serious deterioration in the permeability. If the compaction is carried out with a bulldozer, which does not cause such remoulding damage, then the permeability is not seriously affected. Compaction is essential in order to prevent differential settlement.

In the hydraulic construction industry colliery shale is (in the Netherlands) commonly delivered in two gradations, i.e. 0–70 mm and 10–125 mm. The volumetric mass of the particles is about 2,200 to 2,500 kg/m^3. The volumetric mass of the loose dumped material is 1,700 to 1,800 kg/m^3.

5.4.3 *Silex*

Silex, also known as flint, is a very hard rock, consisting almost entirely of a microcrystalline form of SiO_2. The volumetric mass is about 2,600 kg/m^3. The volumetric mass of the loose-dumped material depends of course on the particle grading and is about 1,600 kg/m^3; this can be increased to about 1,800 kg/m^3 by mechanical compaction.

Silex is found in more or less regular layers (strata) in the limestone deposits (marl) in

Zuid-Limburg (the Netherlands). When winning limestone for use in cement manufacture, part of the silex remains as a by-product. Delivery of silex is in general in three size ranges, i.e. 0–25 mm, 0–90 mm and 25–70 mm.

Because of the production and process method used, the silex always contains some "tauw", which is pure limestone recognisable by the colour. It comes from the uppermost marl strata, which are so hard that it is not crushed in the crusher, but separated out with the silex.

5.4.4 *Slag*

As an underlayer to and filler between revetment blocks, various types of slag are used:

a. Litz-Donawitz (or LD) slag.
   It is a slag produced as a by-product in the LD steel production process. The particles must have a volumetric mass of at least 3,100 kg/m^3.
   LD slag can be contaminated by particles of lime, and encapsulated steel and iron. A disadvantage is that the hydraulic action of the lime can cause petrification.
b. Phosphorous slag.
   This slag is a calcium silicate which is derived from phosphor ore by the addition of gravel (amongst other items). The volumetric mass of this slag is 2,800 kg/m^3. It is angular in shape and looks like natural rock.

Slag from lead ore with a high volumetric mass was often used in the past. However, it is no longer applied because of the danger of lead leaching out which makes it no longer acceptable in the environment.

5.4.5 *Geotextile cloth or membrane*

The strength of a woven material is in general expressed as the force exerted at failure per unit of length in the direction at right angles to the direction of the force. The measured strength also depends on the shape and dimensions of the test strip, the method of clamping and the loading. The reduction in the long-duration strength in comparison to the short-duration load can in some cases be 50 % or more. In civil engineering constructions, the geotextile is generally not subjected to an imposed stress but more to an imposed deformation. Geotextiles are well able to cope with such deformation due to the large permissible extensions, provided that the deformation does not occur over too short a distance.

Deterioration of the mechanical characteristics can be caused by chemical or photochemical attack on the material of the fibres or by mechanical damage to the construction [17].

Geotextile durability is in the first instance dependent on the molecular structure of the artificial fibres. Resistance of the fibre material against attack can be improved by the addition of protective substances to the base material. Carbon can be added to reduce the adverse effects of Ultra Violet (UV) radiation, as well as various anti-oxydants. Research, however, has shown that anti-oxydants in the long run leach out to a substantial degree. Iron ions will shorten the life of polypropylene fabrics. Acids can have negative effects on the action of anti-oxydants.

Woven polyamids (nylon) are not subject to the ageing process to any notable degree.

5.4.6 *Sand asphalt, bitumenised sand*

Sand asphalt comprises sand, filler and bitumen, whilst bitumenised sand contains only sand and bitumen. The bitumen content is 3 to 5 % (by mass), which makes it open-structured as the bitumen only serves to bind the sand particles together.

The small bitumen quantity is just sufficient to cover the sand particles with a thin bitumen film of a few microns thickness. Moreover, the bitumen concentrates on the areas where the sand particles touch. This means that bitumenised sand, depending on the degree of compaction, the particle distribution and particle shape, has a large permeability which approaches (for the practically possible compaction) the permeability of cohesionless sand.

Durability is determined by the durable binding qualities of bitumen.

For a more extensive treatment of this subject reference is made to [33]: "The use of asphalt in hydraulic engineering".

5.5 **Filter characteristics**

5.5.1 *General*

For more background-information, reference is made to [15], [16], [17] and [18]. The most important requirement to be fulfilled by a filter in slope protection works is its ability to protect the subsoil against wash-out as a result of flows caused by waves or a hydraulic gradient. The filter is to prevent migration of sand particles. In considering the sandtightness of filter constructions, two conditions must be distinguished:

– sandtightness independent of flow conditions, no matter how strong the flow is;
– sandtightness under the condition that specified limits in the flow conditions are not exceeded.

With the sandtightness independent of the flow conditions, the tightness depends on the fact that particles of the subsoil cannot penetrate into the filter material due to the particle size of the subsoil being greater than the pore dimensions in the filter.

With sandtightness dependent on the flow conditions, it is important to know which direction the flow has in relation to the filter orientation.

If the flow is at right angles to the interface between filter and base material, the filter will always be sandtight if the hydraulic gradient in the base material is smaller than about 1. The gradient force is in equilibrium with the weight of a column of sand (specific weight of 2,650 kg/m^3 and a pore number of 0.4). The gradient of 1 is also known as the "critical" or "fluidisation" gradient.

Without cohesive forces between the particles, or other additional pressures due to surcharges, the equilibrium will be disturbed with gradients in excess of 1 and movement of sand particles is possible. With such steep gradients it is important to know whether the flow changes direction. If the flow does not change direction, the soil particles can "arch" at the entrance to filter pores, which improves the sandtightness. Under such circumstances a "natural" filter can be formed also, so that for sandtightness it would be sufficient to exclude only the largest particles from migration through the filter. During the formation of a natural filter, all particle transport ceases because the remaining coarse fraction functions as a filter for the underlying layers. Under a cyclic loading a natural filter could be destroyed.

If the flow is parallel to the separation plane between filter and base material, the items of major importance for sandtightness are the hydraulic gradient and the thereby induced flow velocities within the filterlayer. When a critical value of the gradient is exceeded, the flow velocity in the filter could become so great that the base material begins to move and the sandtightness will be lost.

For granular filters, the internal stability of the graded mixture is important. For geotextile filters, attention should be given to blockage, clogging by silt particles, strength of the cloth (or membrane), elasticity and durability.

5.5.2 *Types of filters*

Granular filters

Historically, filters have been used which consisted of granular materials, which can be coarse, fine, rounded, flat or angular, more or less well graded, with a large or small volumetric mass.

A filter of that type can comprise one layer with a nearly homogeneous particle gradation, or comprise several layers with a gradually increasing particle size.

An advantage of the granular filter is its easy adaptability to the filter qualifications by virtue of the large degree of freedom in the composition of the granular mixtures. However, that advantage (and freedom) is limited by financial considerations.

Geotextile filters

The most common forms of geotextiles are:

- mesh fabrics;
- ribbon fabrics;
- mats;
- cloths;
- membranes.

Mesh fabrics are woven using nearly cylindrical threads (monofilament). Characteristic of mesh fabric is the regular pattern of openings and the large percentage of openings per surface unit. The size of the openings is mainly governed by the filament thickness and the number of filaments per unit length.*

Ribbon fabrics are woven with artificial fibre ribbons lying flat and tightly together in the fabric. Characteristic of this fabric is the very small percentage of openings per surface unit.

Mats are woven of split film filaments, made of strengthened film, fibrilated or not, and possibly twisted. The filaments are through this process turned into a fibrous structure. The size of the openings depends mainly on the thickness of the filaments and their spacing.

Cloths are woven with multifilament threads, twined or non-twined. The threads are packed closely together. Because the weave is thin, it remains pliable, strongly resembling a textile.

Membranes comprise long or short fibres, which possess cohesion with or without a binding agent. Characteristic of this material is that it looks untidy but very dense. Layer thickness can vary from several millimetres to fractions of mm.

Geotextile filters have the advantage of being very thin, especially when compared to multi-layer granular filters, but they are on the other hand very easily damaged.

Composite filters

In this group, a distinction can be made between granular filters with a binding agent and granular filters wrapped in geotextiles.

Examples of the first group are the open-textured mixtures of sand and stony materials, bound with bitumen, such as sand asphalt, bitumenised sand and open stone asphalt. The bitumen provides the binding agent, which makes the granular filter more resistant to external loadings.

5.5.3 *Sandtightness requirements*

Granular filters

a. Independent of the flow conditions for nearly uniform filter and base materials (i.e. very steep sieve curves), sandtightness will be satisfied if (see Fig. 10):

$$D_{50f} \leq 4 \text{ to } 5 \ D_{50b} \tag{1}$$

where D_{50} is the diameter of the sieve aperture through which 50 % of the sample passes. Index f=filter and index b=base material.

For filter and base materials with non-uniform particles (relatively flat sieve curves, see Fig. 11), the criterion is:

$$D_{15f} \leq 4 \text{ to } 5 \ D_{85b} \tag{2}$$

---

* Rankilor in "Membranes in Ground Engineering" (Published J. LIZLEY) speaks of the "number of picks per cm".

b. Depending on the flow conditions, it is sometimes possible to relax the requirements set out under (a) above. It is, in those circumstances, recommended that the advice of a specialist research laboratory is obtained.

It is moreover desirable to aim for the sieve curves of filter and base materials to be as parallel as possible (Figs. 10 and 11), because the validity of formulae (1) and (2) depends thereon. Compaction of filter material has to be carried out in layers which are not too thick.

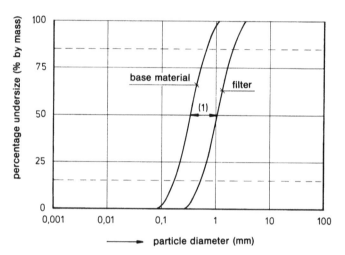

Fig. 10.   Filter requirements for uniform material.

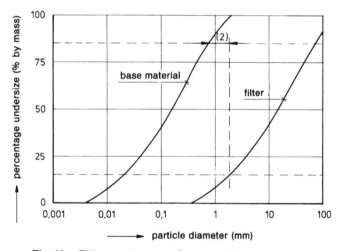

Fig. 11.   Filter requirements for non-uniform material.

Geotextile filters

a. Independent of the flow conditions, sandtightness is achieved if:

$$O_{max} \leqq D_{15b} \tag{3}$$

where O_{max} is the largest aperture in the geotextile filter. In practice, O_{max} is usually given the value which is the equivalent of the average diameter of those sand particles belonging to the 2 % (by mass) which passes through the filter: O_{98}.

b. Dependent on the flow conditions.

When a "natural" filter is formed through the action of cyclical loading, the following sandtightness condition is valid:

$$O_{max} \leqq D_{85b} \tag{4}$$

Under steady flow conditions, for mats, mesh fabrics, ribbon fabrics and cloths the formula is:

$$O_{90} \leqq D_{90b} \tag{5}$$

For membranes the limit is more favourable:

$$O_{90} \leq 1.8 \ D_{90b} \tag{6}$$

For the design of filters the first consideration must be to investigate the flow conditions. If these are strongly cyclic with hydraulic gradients in the base material greater than 1, the filter rules of case (a) above should be used. For other conditions of flow the rules under (b) can be employed.

When flows are mainly parallel to the separation plane between filter and base material and the hydraulic gradient is not too large, even very open filters could be sufficiently sandtight.

Revetment constructions for coastal defences are often subjected to strong cyclic flows and, therefore, need to satisfy the rules indicated under (a) above.

5.5.4 *Other requirements*

The permeability of a filter also has to satisfy the requirements related to the danger of uplift of the revetment due to porewater pressures within the bank. On the other hand, a low permeability filter has the advantage of providing a more stable base for blocks under wave attack.

For the internal stability of granular filters the following general rule can be applied:

$$D_{60f} \leqq 10 \ D_{10f} \tag{7}$$

which assumes that no internal migration will occur, irrespective of the steepness of the hydraulic gradient.

It should be noted that for small hydraulic gradients this criterion can be eased. However, no accurate values are known at present for the critical hydraulic gradients relating to internal instability.

In order to prevent blockage of a granular filter, the following rule applies:

$$D_{5f} \geqq 75 \ \mu m \qquad\qquad (8)$$

5.6 Quality control

For an embankment to function efficiently for a long period of time, certain requirements will have to be satisfied by the construction as a whole, as well as in parts, and, therefore, the construction materials should also meet certain specifications. In the foregoing discussion, various pertinent characteristics of construction elements and materials have been described and in some cases numerical values have been given. In order to ensure that the materials delivered to the construction site (and consequently the structure) permanently meet the requirements, it is necessary to provide quality control. This means that the project design includes which material characteristics are essential and what quality level will be required. The requirements have to be formulated in a way which makes control possible. It is also necessary to have test methods available, or to develop them. It is then also necessary to investigate whether the material as required can be produced and placed on site, and what effect the site construction work has on the characteristics of the material. Finally, it has to be decided which quality checks have to be made on delivery and/or during construction and what the consequences might be of that procedure.

Naturally, the procedure described above is inevitably based on feedback to and adjustment of the original starting-points. A situation where by requirements and test procedures are absorbed in standards, codes and guidelines, such as is the case for concrete (Chapter 4), has not yet been reached for the various hydraulic construction materials. Developments are in progress, with (often interim) reports available. Much information presented in the preceding sections was derived from such reports.

It is advisable that use is made of the experience and knowledge available with the specialist organisations on quality control.

The ultimate advantages of a consistent quality control lie not only in the qualitative improvement of constructions and greater understanding between supplier and client, but also in a more constant and/or improved quality of materials. This provides economy in construction and makes it possible to provide more sophisticated designs.

In carrying out quality control the following stages exist: preliminary investigation, investigation on delivery and/or during construction and checking the final construction. The preliminary investigation is directed at the determination of the suitability of a particular material or mixture (of materials) for the proposed application, also with regard to the availability of the required quantities. All this is based on the proposed requirements (whether or not included in the codes). Investigations on delivery and/or during construction include checking the desired quality, with the main purpose of maintaining a constant quality level (homogeneous material). Checking the final construction is aimed at the quality of the final total result as well as the quality of all the constituent elements.

Especially the checking on delivery as well as the final construction will determine any rejection, replacement, discount and guarantee.

It depends on the materials used and their application in which phase of the quality control process the main checks are made. This also applies to where the checking takes place (in a factory or on site).

The responsibility for the delivered quality (of products or materials) lies with the supplier, producer or contractor. The employer checks on the quality through specified controls, carried out by or for him, or through information on tests carried out on behalf of the supplier, the operational control.

For the materials mentioned in Section 5.4 (colliery shale, silex, gravel and slag), the filter requirements are expressed in particle size distribution limitations (see Section 5.5), which have to be checked by means of sieving at source or on delivery (sieve test). Where the particle size distribution (and consequently the filter characteristics) can alter due to mechanical or other attack, investigations have to be carried out on durability and strength (frost-thaw test, boiling test, determination of alkaline/lime content, crushing test, etc.). In some cases the permeability/density is determined in relation to flow stability or in connection with strength characteristics.

The determination of the percentage contents of sand, organic matter and clay fraction in clay has to be carried out in connection with the expected performance of the material, but also because of the importance of its homogeneity. Investigation of the moisture content is of major importance for workability during construction. Erosion and shrinkage characteristics need to be investigated in the preliminary investigation.

Geotextile filters are checked for permeability and sandtightness. Strength and flexibility (elasticity) are of importance during construction and in connection with settlements in the service stage; various shecks exist to determine these parameters. The filter material is important for the life expectancy. For polypropylene there is for example an accelerated ageing test. As the geotextiles are produced in factories, quality control should preferably be carried out in the factory.

Bitumenised sand needs special attention on the percentage content and the type of bitumen used, and this should be determined at the production stage. Supervision of transport and placing on site (temperature, placing method) is important to ensure that the required layer thickness and homogeneity are obtained.

CHAPTER 6

LOADING ZONES ON EMBANKMENTS

6.1 Sea walls

The degree of wave attack on a sea wall during a storm tide depends on the orientation in relation to the principal direction of the storm, the duration and strength of the wind field and the extent of the water surface fronting the sea wall.

The presence of a foreland, i.e. saltings or sand banks, and the height and width thereof, or deep channels and the width of these channels, or the presence of harbour break-waters, all have their influence. High foreland will reduce the local wave attack on the dike to a greater or lesser degree. If the sand banks are located some kilometres away from the embankment, with appreciable depths in the area between sea wall and sand-bank, then the initially reduced waves can grow higher again. Thus refraction and diffraction will, in the North Sea for example, change the character of the waves substantially when they approach the deep channels and sand shoals near the (Dutch) coast. The wave motion further into estuaries is not solely determined by the sea waves. Tidal currents along the sea wall also have an effect on the wave action. Propagation of wave systems can to some extent be calculated provided that the area concerned is topographically not too complicated. Such calculations should, where possible, be supplemented and controlled by site observation (marks left by floating debris, use of aerial photography).

Because the process of wave distortion is so complicated, it is mostly not possible to calculate more than an approximation to the wave height at the dike.

There is a certain correlation between the water level (tide plus wind set-up) and the height of the waves, because wind set-up and waves are both caused by wind. For sea walls in the tidal region, fronting deep water, the following approximate zones can be distinguished (see Fig. 12):

I. The zone which is permanently submerged.
II. The zone between MLW (mean low water) and MHW (mean high water). This is the zone where wave attack occurs daily. The ever-present wave loading is of importance, although the wave height is mostly less than in Zone III.
III. The zone between MHW and the design level. This zone can be heavily attacked by waves, but the frequency of such attack reduces as one goes higher up the dike.
IV. The zone above design level, where only wave run-up occurs.

For sea walls with a high-level "foreland", Zone I is not applicable. Depending on the level of the foreland, below or a little above MHW, Zone II may exist partially or not at all.

48

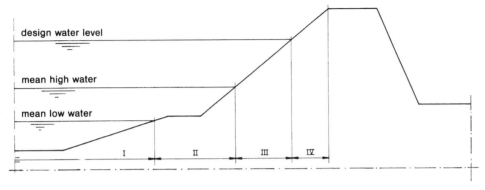

Fig. 12.   Division of the slope into zones on sea walls in the tidal region without a foreland.

Embankments in the tidal area with a high foreland have the benefit of the wave-damping effect of that foreland, making wave attack at lower levels less heavy than on dikes lying along deep water.

A bank slope revetment in principle functions no differently under normal circumstances than under extreme conditions. The accent is, however, more on the persistent character of the wave attack rather than on its size (see Fig. 13). The quality of the seaward side slope can, prior to the occurrence of the extreme situation, already be damaged during relatively normal conditions to such a degree that its strength is no longer sufficient.

The division of the slope into loading zones has a direct connection with the safety against failure of the revetment; for this subject reference is made to Chapter 14.

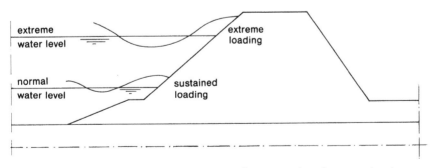

Fig. 13.   Difference in intensity and location of wave attack under normal and extreme conditions.

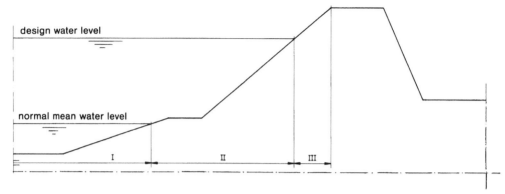

Fig. 14.   Division of the slope into zones for lake dikes.

6.2   Embankments along lakes

Where the water level against a sea wall is changing continuously under the influence of tidal movement, the wave attack will very rarely persist for any long period at the same level. Along lake embankments, in contrast, there will (apart from exceptional circumstances) be little variation in water level*, and the daily wave attack will occur at approximately the same level; i.e. the average water level.

Lake embankments without a high foreland have the following approximate zone distinctions (Fig. 14):

I.   The zone which is permanently submerged. This zone has to be protected against wave attack and currents, by means of a construction to be laid and maintained below water level. Placing of blocks is not possible in this case. A different situation can occur in a closed-off sea-inlet where the average water level lies above the original low-water level, at a level some way up the revetment slope. This has important adverse consequences for the maintenance of the slope below water level.

II. The zone between the normal mean water level and the design level. This zone is attacked by waves and currents, ice and other floating debris. In this zone, waves break just below the actual water level.

III. The zone above design level, where only wave run-up occurs. A grass cover is often adequate to meet the requirements in this zone.

6.3   River embankments

River embankments are generally constructed as green dikes. Only where the embankments are exposed to serious attack, a hard revetment is applied:

---

* This is not necessarily valid for water storage lakes where wave attack at draw-down is at very different levels.

50

- at locations where the channel cross-section narrows;
- at locations where the main current runs close to the toe of the bank slope, e.g. the outside of bends;
- at locations with an unfavourable position with regard to the wind;
- in the proximity of locations with discontinuities in the cross-section, such as bridge piers, etc.;
- where the dike slope is flanking deep water.

The locations where revetments have been constructed were generally determined on the basis of experience of the controlling authorities, following on the history of damages in the past. Such revetments are usually loaded only by currents; exceptionally, in cases where there happens to be a long wind fetch, wave attack can also be expected. At some locations additional attack can be created by waves and drag velocities due to passing ships.

In contrast to the situation with sea walls, river embankments often have to hold high water levels during prolonged floods (lasting for weeks). Under those conditions, the dikes can become saturated. When flood waters fall again quickly (within a period of a week) the porewater in the dike must be able to drain out freely. This means that revetments should be of an "open" structure in comparison with the permeability of the layer under the revetment.

CHAPTER 7

EMBANKMENT PROFILE

The shape of the bank slope needs to be observed longitudinally as well as in cross-section.

7.1 Cross-section

Important aspects in the choice of the cross-section include, amongst others:

- the slope angle or gradient;
- the shape of the cross-section;

Gradient of the bank

The gradient may not be so steep that the whole slope or the revetment can lose stability (through sliding), both during construction as well as in the service stage. These criteria give, therefore, the maximum possible slope angle. A flatter, more gentle, slope leads to a reduced wave force on the revetment and less wave run-up; wave energy is dissipated over a greater length. Using the wave run-up offers, for example, the opportunity to calculate the crest height of a trapezoidal profile of a dike and thus determine the volume of the embankment at a given crest width.

However, this does not necessarily imply that minimum earth volume coincides with minimum costs. An expensive part of the embankment comprises the revetment of the waterside slope and the slope surface increases as the slope angle decreases. Similar considerations hold for dikes with a sea-side berm.

The optimum cross-section (based on cost) can be determined when the cost of earth works per m^3 and those of revetment per m^2 are known. Careful attention is, however, needed because the revetment costs are not independent of the slope angle.

Another point to be taken into account in the choice of a slope angle is the space occupied by the embankment; this could be the decisive factor in case of the presence of buildings compared to the before-mentioned economic optimization.

Shape of the cross-section

The slope can be flat, convex or concave; various opinions exist on these alternatives. With a concave profile the wave run-up is reduced at lower water levels compared to run-up on a convex slope, but the converse is true at higher water levels. The flexibility of the block revetment, in connection with possible settlement and soil erosion, will be adversely affected on a convex slope, due to the arching action. A concave profile has in this respect an advantage over a convex profile.

It is not easy to determine the value of the advantage or disadvantage of the flexibility as a function of the shape and degree of curved slopes. The concave shape makes it more difficult for wave action to remove individual blocks, because of the "upside-down" arching effect of the concave slope. A disadvantage of a pronounced convex profile is that with many systems the block joints will become much too open.

The water-side berm is an element in dike construction, of which the function has changed in the course of time. Particularly in Zeeland* outer berms have been applied frequently. It could in the past lead to a reduction in the expenditure on stone revetments. On a very gently sloping berm a good grass cover can be maintained, even at lower levels, better than on the steeper angle of an uninterrupted slope. Moreover, the outer berm produced an appreciable reduction in wave run-up.

The condition for a satisfactory realisation of the two main functions, i.e. economies on stone revetment and reduction of the wave run-up, is that the berm level has been adjusted correctly to the normative water levels which can occur in front of the embankment. It will be obvious that the berm level should be higher if the dike has a less sheltered location; in different cases there can be differences of some metres in height. Furthermore, the outer berm level, in order to obtain a substantial reduction in wave run-up, should be close to the still-water level of the design storm surge. If the berm lies too much below that level, the highest storm surge waves would not break on the berm and the run-up will be inadequately affected.

Present practice is to place the outer berm at design water level as indicated in the Delta Commission report. For a storm surge berm at such a design level there are in general no problems with the growth of grass on the berm and the upper slope. However, there can be circumstances which require also the application of a hard revetment on the berm and even on a part of the upper slope (see fig. 15).

In order to avoid the action of extra forces at corners in the profile of the outer slope, all such corners should be rounded as much as possible. An important function of the sea-side berm is its use as an access road for bank maintenance.

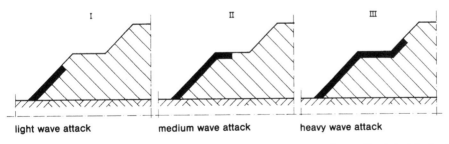

Fig. 15. Possible termination of the hard revetment at an outer berm situated at design level.

* South-west Province, comprising mainly islands.

53

River (flood) dikes often have a revetment starting high up on the slope. It then needs solid support along the bottom line. Care should be taken to prevent erosion of the grass cover at that junction with the revetment.

7.2 Longitudinal profile

Due to irregularities in the longitudinal profile of an embankment, in connection with the topography of the terrain in front of the dike, some reaches of the slopes could be subject to more than normal attack (for example due to refraction of waves).
Not all revetment systems are suitable for use on a curved slope, due to several complications:

- systems which do not permit deviations from a straight line;
- going around curves leaves gaping joints;
- difficulties in placing the blocks by machine.

Mechanical methods for placing the blocks is in practice limited mainly to straight lines and to large radius bends with sufficiently large areas. Placing of blocks by mechanical means is not only economical, but it can also clamp the blocks tightly together, much tighter than can be achieved manually. This contributes to better interlocking elements.
It is, however, important to remember that any damage to a block revetment should be repairable manually.

CHAPTER 8

BOUNDARIES AND TRANSITIONS

The experience of many controlling authorities is that much damage takes place at transitions from one type of revetment to another and at the line where the revetment ends. It is possible to give a great deal of care and attention to the revetment construction as such, but when the weakest link breaks, it will be normative to the safety of the defence system. This chapter is, therefore, dedicated to these special aspects.

8.1 Toe construction of revetments

The toe construction has the function of revetment support and to protect it against erosive action on the bottom edge of the slope.

A distinction has to be made between a sea wall and a lake embankment. The daily occurring wave attack on a lake dike, in contrast to that on a sea wall, will be approximately at a constant* level. The daily wave attack will have to be countered through that part of the revetment where the underwater slope meets the revetment on the slope above water.

On sea walls the toe construction is usually, for constructional reasons, placed above low water. Wave attack of any significance will not often occur at that level, because the winds causing waves to occur generally also raise water levels, so that even at the official low water time, the waves will occur at a higher level. An exception to this is the sea swell which can indeed give heavy wave attack at lower levels.

Where embankments have high-level forelands, the toe of the dikes are protected against the heavier waves by the breaking of waves on the high ground levels. Where a slope has to be revetted with rectangular blocks, a need is created for "straight lines". Such lines are provided by means of boards fixed to a row of timber piles. The boards could be either of timber or concrete. This method of construction is certainly not always adequate for heavily attacked lake embankment toe constructions. Even if there is a support berm in front of the toe, with a heavy mattress, there is still a chance that the wave motion will gradually remove sand to the base level of the boards, so that the foundation under the lowest revetment blocks can be washed out and the toe will subside.

An improved construction in this case would comprise a sheet pile wall of tongued and grooved boards, preferably strengthened along the top with a waling (see Fig. 16). This construction provides the toe with a more gradual transition from the slope to the level

---

* Except for water storage reservoirs where water levels can vary greatly, as stated previously.

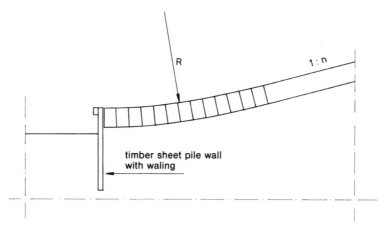

Fig. 16. Toe structure with timber sheet pile wall as a possible solution for lake dikes.

of the stone-protected berm, which is an important aspect considering the "to-and-fro" wave motion. This system also provides a better means of connection with the first row of toe blocks.

Design of a toe construction requires due attention to the following aspects:

a. If the blocks are placed directly against a solid line of timber posts, the danger of wash-out of the base material exists because of unavoidable gaps between the posts; moreover, this method does not produce straight lines.
   The starting-point for rectangular blocks has to be the toe board. Exceptionally, the solid line of posts could be considered for use with polygonal blocks, but great care will have to be given to the sandtightness of the toe construction.
b. If the toe construction projects too far above the adjoining foreland or support berm, there will be a chance of slope undermining, with the toe construction "toppling over"; a possibility which will usually manifest itself during the construction stage.
c. If the dumped stone level on the support berm is pitched too high, the transition between berm and toe will be discontinuous which, especially in case of receding waves, leads to awkward extra forces acting on the toe construction. The stability of the support berm is also affected.
d. In case the toe board is not sufficiently deep, the slope could be undermined and, with reduction of foreland or berm level, the stability may be lost.
e. With a weak subsoil there is a danger that a sheet pile wall or a row of piles may not be sufficiently stable.

8.2 Top edge of the hard revetment

The transition from the hard revetment to the grass cover often forms a weak point in the construction. A good grass cover is capable of resisting breaking waves to some

56

degree, but these must not be large nor occur too frequently. The more frequent wave forces and run-up have to be absorbed by the hard revetment.

The location of the revetment/grass cover transition can theoretically be selected at a lower level if account is taken of the wave run-up limiting effect of a rough and open slope, the presence of an outer berm, the wave-reducing effect of a high-level foreland, or a combination of all these possibilities. The determination of the transition location is, however, often a matter of experience combined with local circumstances (debris-line observations related to past occurences of damage; the "trial and error" method) or the application of a practical "rule of thumb" (for example half-way between design level and bank crest level).

Knowledge on the resistance of grass covers against breaking waves and wave run-up, coupled with the magnitude and frequency of occurrence, is at present limited. At a transition, care has to be taken that the discontinuity is as small as possible in order to prevent the possibility of local undermining. It is, for that reason, not recommended to drive a solid line of posts, along the revetment's top edge, which projects above the plane of the revetment (as is sometimes done to prevent floating debris from reaching the grass cover!)

The principle of the transition construction must be: a greater strength than the grass cover with a hydraulic roughness about equal to that of the grass cover. A strip of double-thickness "road bricks", placed "on edge" (0.10 m high), with top soil filled joints, would be satisfactory; this strip of bricks must, however, have the opportunity to be thoroughly integrated with a well-grown grass cover.

A less labour-intensive and more effective possibility comprises the use of concrete blocks with earth-filled perforations for grass to grow through. The units used for this type of application are much bigger and heavier than road bricks. Also, the discontinuity of the transition from revetment to subsoil is reduced, and it is to be expected that the grass will form better and deeper roots into the clay subsoil. Nevertheless the transition will remain vulnerable as long as the grass cover is still in the process of developing. Problems can be created when mowing the grass if and when concrete elements are not placed to accurate lines and levels, or when they have settled to different levels.

When the upper edge is only protected by a concrete strip, a relatively small amount of undermining could already cause substantial damage. In such cases it is recommended that the concrete strip is supported by posts driven to top-of-concrete or a little below that level.

When hollow concrete blocks for grass growth are employed, no support posts are needed as the blocks are sufficiently stable. That construction method has an important additional advantage when the embankments are grazed by sheep, because the sheep's habit of forming a path along the revetment top edge will be prevented.

8.3 Transitions to other revetments

In case two different types of revetment adjoin, special attention should be given to the subsoil.

If, for example, a fine-grained soil exists immediately alongside a coarser material (see Fig. 17), there is a danger that the finer material will penetrate into the coarse material under the influence of groundwater flow and soil pressures. This is detrimental to the drainage function of the coarse material and can also lead to local slope settlements. The "sandtightness" requirement can be satisfied through an effective granular composition of the coarse material (granular filter design). It is often easier and cheaper to use a geotextile as a filter, or use a concrete separator to prevent migration of the finer material.

Problems can often arise at the transitions between coarse materials, for example colliery shale, slag, gravel, waste, etc., and the finer soil such as sand and clay. The situation shown in Fig. 17, where the penetration of the fines into the coarse materials is indicated, also needs to be considered in terms of measures required in order to prevent undermining of the asphaltic concrete when damage occurs to the concrete block revetment.

Various methods of construction to reinforce the transition are possible. For example the following:

a. Let the asphaltic concrete extend a short distance over the colliery shale (instead of the sharp break shown in Fig. 17). A disadvantage is that wave impact pressures will be transmitted underneath the asphaltic concrete as well, with the associated danger that the asphalt cover is lifted.

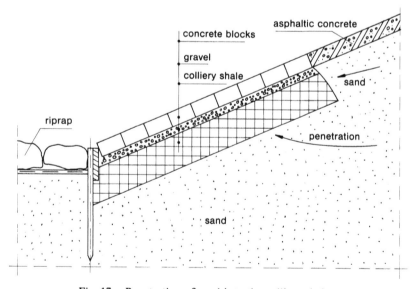

Fig. 17. Penetration of sand into the colliery shale.

b. Increase the thickness of the asphaltic concrete in a strip along the joint. This can, however, cause problems during construction, where a trench has to be cut to provide the depth required.

c. Construct a row of posts with boards (keeping tips just below the slope surface).

d. Fill the joints of the upper layers of blocks with bitumen. This is only possible if the joints offer adequate space to do so.

It is possible with the application of an edge of concrete strips (see Fig. 18), at the transition from concrete blocks to basalt columns, to prevent the clay penetrating into the waste base layer under the basalt columns. The concrete edging should be sufficiently deep then.

The experience of the authorities in charge of embankment maintenance shows that considerable damage occurs at transitions such as that shown in Fig. 18. A possible cause thereof may be that the basalt columns are not held sufficiently tight by the concrete strip, and also the foundation discontinuity at the transition. With the basalt blocks thus being less tightly packed, and the impossibility of increasing wave-impact pressures in the waste foundation layer escaping into the clay under the concrete blocks, the basalt blocks would be subjected to greater uplift pressures. Greater stability can be obtained by penetrating a strip of approximately 0.5 m of basalt blocks along the transition line with liquid bitumen.

For concrete slope revetment systems which have polygonal shaped elements, similar to the basalt block shape, and which also have to be placed on a granular filter, no special attention needs to be given to the transition. The polygonal shaped concrete columns can be placed directly onto the existing basalt slope. Care is needed in one

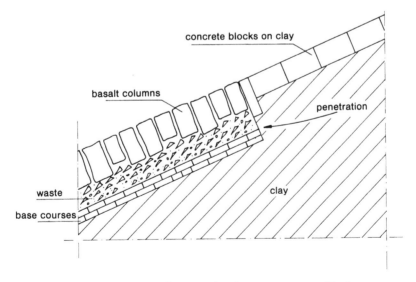

Fig. 18.   Transition from basalt columns to concrete blocks.

respect, i.e. that the granular filter material under the concrete block cannot migrate into the waste layer under the basalt, because it could lead to subsidence. In such cases a granular transition filter will be required.

In Fig. 19 illustrates to what extent the pressures under a revetment can rise, for instance as a result of the closure of a permeable filter with a concrete board. The flow nets shown were determined by computer. For more information on this subject reference is made to the CUR/COW report "Background to the guide to concrete dike revetments" [32].

It is recommended that, as a general rule, transitions to other types of revetment in the longitudinal direction of the embankment are avoided as much as possible. It is

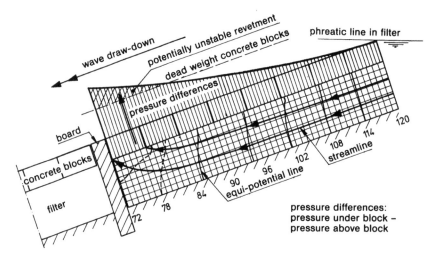

a. pressure differences due to board in filter

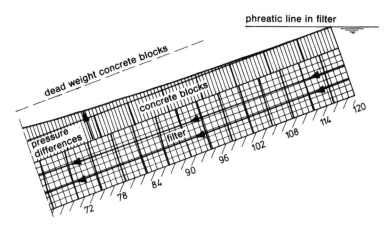

b. pressure differences for continuous filter

Fig. 19.   Increased pressures associated with disturbances in the permeable filter.

60

advisable that a dike slope is protected by one type of revetment from the toe to the top edge, and especially has an underlayer with the same permeability throughout. This is often, however, not possible. For instance when an existing revetment is being extended, and, as is the case with modern embankments, at the transition of a hard revetment (stone) to an asphaltic revetment.

In case concrete edge strips or boards are placed at right angles to the longitudinal bank direction, the matter is more favourable. In these cases only those waves which arrive on the slope at an angle could cause some additional uplift pressures under the blocks in the corners along the top edge. An advantage of concrete edge strips placed in this direction is that the slope is then divided into separate compartments, thus preventing steady progression of any damage along great lengths of the dike.

In order to use concrete edge strips as stops to further damage development, these edgings have to be sufficiently heavy and deep. Making edgings too deep can be dangerous as it would make a separation cut in the foundation layer.

It is very important to ensure that blocks form a good and strong fit against the concrete edgings.

CHAPTER 9

CONSTRUCTION ASPECTS

9.1 General

The concrete block revetment differs mainly from a natural stone revetment by being more regular in shape and also more accurate in measurement. This type of revetment can, therefore, be placed in the traditional "manual" method, using specialised labour (stoneworkers), but also by means of machines operated by skilled men. Using cranes fitted with special clamps the blocks can be picked up and placed quickly and accurately, which is particularly useful for the larger size blocks. The increased production rates are often required as most site construction works on coastal defences have to be carried out (in Holland) between mid April and mid October, and even then probably only when tides permit access.

9.2 Product

The blocks, columns or slabs for concrete revetments should be manufactured in well-equipped factories under expert supervision in order to ensure a constant high-quality standard, both for the concrete composition (particularly important for the aggressive coastal environment) and for the dimensional accuracy needed for mechanical placing. All appropriate revetment elements should be quality tested according to the standards specified (see Dutch Standard NEN 7024 and Chapter 4).

9.3 Storage and transport

After fabrication the blocks have to be stored for hardening and mechanical damage has to be prevented. Handling operations should therefore be limited to a minimum.
For transport to the site, the storage area must be easily accessible by road or water. As it will be very rare that blocks delivered from factory storage to site can be placed directly on the slope, in view of construction in the tidal zone and the lack of access roads, it is generally necessary to have a temporary storage yard on site.
When blocks are destined to be placed mechanically, the units should be appropriately assembled at the factory so that they need no rearrangement for on-site placement. Site stacking should be done on level ground to prevent the stacks sliding or falling over. It is important to check that storage and transport are arranged in such a way to eliminate soil adhering to the units as this would cause problems with accuracy of placing the blocks. On-site transport should be carried out with vehicles suited to the terrain.

9.4 Construction

Revetments comprising concrete blocks or columns are in general placed directly on the prepared slope materials, e.g. clay, colliery shale, broken stone (gravel size) or silex. Because the concrete blocks are dimensionally accurate, it is necessary to finish the base accurately to profile. Templates are in general use for this purpose, with planers or scrapers employed to produce the desired profile and levels, depending on the base material used, e.g. clay, filters, etc. (see Chapter 5).

Placing larger units by hand requires much labour and limits production, so that this method will be only used to obtain good transitions to existing revetments, in case of repairs, or in case of systems which are not suited to mechanical placing.

For mechanical placing, cranes are used to pick up several elements at once (with clamps) for direct placing on the slope. These multiples of elements can cover areas between 1 and 3 m^2. These can be placed by the crane with some force to provide a good connection with the adjoining blocks. Production can of course be increased by employing several cranes.

Revetments with large joint spaces, as in the case of polygonal concrete columns, need to have the joints filled with great care, for example with broken stone, to produce a firm, immovable, but permeable surface.

Problems do occasionally occur with the joint filling in the latter case, when blowing sand fills the joints before the broken stone can be hemmed in.

CHAPTER 10

MANAGEMENT AND MAINTENANCE

10.1 Inspections

There is at least one annual inspection of the complete slope protection, primarily for safety reasons, carried out by those who are responsible for the maintenance.

The responsible bodies are commonly taken to be the district "Water Boards"; Provincial and Central Government embankments also exist as do occasionally local Council embankments.

Attention is directed towards essential repairs of, for example, subsided parts of the slope and renewal of worn-out areas, or extension of existing slope protection. After each storm tide, such an extensive inspection is conducted at low tide to locate any possible damage. (This is usually done by the engineer or surveyor to the Board). Maintenance staff, employed by such Boards, take note of any damage during their daily work schedules.

As the Provincial Governments are usually charged with the task of overall supervision of coastal defences, the Provincial technical staff also carry out inspections. Due to the supervisory task, the accent is on the control aspect of inspections rather than on the duty to find damaged dike slopes.

10.2 Maintenance

Annual maintenance covers broadly the following activities:

– Repair of damage after heavy wave attack. This can be limited to very local damage, for example a single block that was "lifted" and not returned to its own position. It may, however, also concern more extensive damage, for example at connections to other constructions.

Depending on the nature of the damage, the repair would comprise the breaking out of a single element and replacing it with a fresh concrete fill placed on site. Alternatively, a greater area has to be broken out, and after filling the underlayer to the correct levels, the blocks or new elements are put back in again.

Where damage occurs repeatedly at the same location, investigations would clearly be required into the construction of the slope in that area. Hence appropriate countermeasures can be taken.

– Repairs on site of subsidence which occurred as a result of soil losses or due to collapse (displacement) of the toe construction. In these cases the toe construction and the underlying materials will be improved at the same time. Costs are often high because the whole slope revetment has to be removed from the toe up to the top edge and on both sides at an angle of about 45° to the top.

- Extension of existing revetment areas in an upward or downward direction can also be counted as part of the maintenance. Downward extension in case some of the foreland is removed, and upward extension if the damage pattern indicates the need for extended protection.
- Improvement works of a limited extent, or works of a greater extent over a number of years, can also be counted as part of maintenance. These improvement works can be fairly radical and include renewal of toe construction, underlayers and revetment construction.
- Repair of damage created by anglers lifting out blocks or due to forms of vandalism.
- Repair of damage due to stranding of ships, deck loads or drifting ice.

10.3 Repair feasibility

In comparison with natural stone revetments, the feasibility of repairing concrete block revetments is in some cases more restricted, because:

- the inter-connection between blocks is often much greater, which makes removal of single elements more difficult;
- it is impossible to push single elements back in their own place in the revetment; it never fits. Moreover, it also generally strains the true alignment.

These limitations are not valid for concrete elements with shapes similar to basalt columns.
Apart from the restrictions indicated above, repair possibilities include the following:

- It is possible to cast some elements in-situ, as is the rule when irregular gaps have to be filled. However, because precast blocks are generally of superior quality, repairs should as much as possible be carried out using precast blocks. If and when in-situ concrete has to be used, attention should be directed to the production of a good and dense concrete and good curing.
- The new blocks needed for the repairs can be delivered readily and in small quantities.

10.4 Reuse

When blocks have been placed without mortar or asphalt joint fillers, reuse of blocks is nearly always possible, unless blocks are too old or of poor quality. With large elements, such as slabs and step slopes, no reuse is possible. Also, for constructions where cement mortar or asphalt was used to penetrate the revetment, reuse is rarely possible (for revetments). It could possibly be used as loose-dumped material.

CHAPTER 11

HYDRAULIC BOUNDARY CONDITIONS

11.1 General

In view of the function of (coastal) water defences, the loads will obviously be mostly due to the actions of long and/or short waves. In broad outline the following wave phenomena can be distinguished:

a. Low-frequency water level changes, such as flood waves, tidal waves, wind set-up gradients and seiches.
b. Wind waves and swell.
c. Ship's waves in navigable waterways:
   - primary wave with the drag-wave forming part thereof;
   - secondary ship's waves;
   - combinations of primary and secondary waves.

These water level variations strongly determine the area which needs to be protected with a hard revetment.

Water level variations on canals and water-storage channels* are comparatively small; probably only caused by lock-water, seepage, drainage and wind effects. Water levels on lakes can vary as a result of wind set-up, inflow or outflow of water, and evaporation. Water levels in a reservoir can change markedly due to filling or emptying, but rainfall and wind set-up can also play a role.

Water levels on a river are determined by the river's discharge regime, and in addition for the lower reaches (estuaries) by tides and also wind set-up. For a coastal defence embankment water levels are governed by tides and winds.

The most complex situation occurs at sea walls, where water level fluctuations can assume many forms. For this reason, and also to reduce the volume of this guide, the considerations here also will be limited to sea walls.

The purpose of this chapter is to provide background information on the various phenomena in preparation for the matters discussed in Chapters 12 and 14. In this context wind waves and the wave deformations in front of and on the banks will be described. For more in-depth considerations reference is made to the CUR/COW report "Background to the guide to concrete dike revetments" [32].

---

* Water-storage channels cover all waterways, in a flat polder area, used for run-off storage as well as for discharge to drainage pumps.

66

11.2 Wave characteristics

11.2.1 *Individual waves*

General

Wave theories have been known for 200 years. These theories are based on the assumption that the wave can by described by the wave height H and the wave period T (or by the wave length L), and refer to regular waves. Regular waves do not occur in reality at sea, but such waves are important because they contain the basic elements of irregular sea waves.

In general, the real wave phenomena are very complicated and difficult to express in mathematical terms, due to non-linearities, three-dimensional characteristics and the random nature of waves. It is the task of the so-called deterministic theories to formulate mathematically as accurately as possible the form of the free surface and the motion of a regular wave for various wave heights and periods and at different water

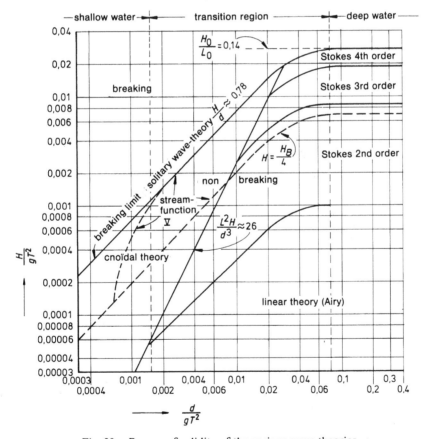

Fig. 20. Ranges of validity of the various wave theories.

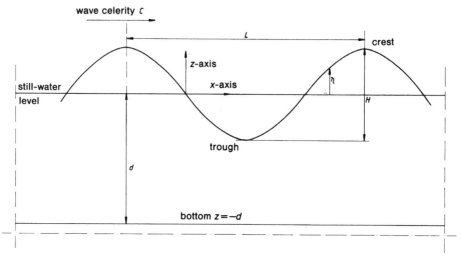

Fig. 21. Profile of a single wave.

depths. In doing so, it is customary to make a distinction between linear and non-linear wave theories depending on which order of flow forces is taken into account.

The significance of the most important theories is presented in Fig. 20. That figure also gives a general view of the validity limits of the various theories. The dimensionless parameters H/gT^2 and d/gT^2 have been used. The variables used are defined as follows (see Fig. 21).

L = wave length: the horizontal distance between two successive wave crests

T = wave period: the time passage recorded at a fixed point between the passing of two successive wave crests

f = wave frequency = $1/T$

C = wave celerity (also known as wave velocity or phase velocity): the speed with which the image of the wave profile travels; it is therefore not the velocity of the water particles

H = wave height: the difference in the levels of the highest and lowest points of the wave profile

d = water depth: measured in relation to the still-water level

There are two classic theories, one developed by AIRY (1845) and the other by STOKES (1880), which describe the regular wave. These theories predict wave behaviour generally better if the ratio of d to L is not too small.

For shallow water the cnoidal wave theory, originally developed by KORTEWEG and DE VRIES (1895), often gives a reasonable approximation for single waves. That theory is, however, hardly amenable to calculation. For waves in very shallow water the best method to describe the waves is the solitary wave theory. In contrast to the cnoidal theory, the solitary wave theory is reasonably easy to use.

68

Sea waves can be subdivided into:

- developing sea, when waves are developing under the influence of the wind;
- swell: these are sea waves which have been withdrawn from the action of the wind (which originally created the waves), either because the wind died down, or because the wind field has moved elsewhere, or due to the waves having moved out of the wind field.

Swell differs appreciably from active sea waves developed under wind influence. Firstly, because the appearance of the waves changes significantly and very quickly after the wind drops or reduces: the white crests have disappeared and the short small waves have been attenuated through internal friction, so that the waves look much smoother. Over longer distances, these waves are subjected to two gradual changes, i.e. loss of height and extensions of the wave period. Because the stability of a concrete block revetment depends strongly on the wave period (see also Chapter 13), the design has to take swell into account in certain cases.

For the prediction of wind waves several methods are available, reference is made to [21] and [24].

Linear wave theory for small amplitudes

Reference is made in this guide to the linear wave theory. A broad outline of this theory will, therefore, be given here.

The linear or Airy-Laplace wave theory departs from the following assumptions:

- sinusoidal surface of the waterlevel;
- small amplitudes, i.e. $H \ll L$, $H \ll d$;
- flat bed;
- "ideal" liquid, i.e. frictionless, incompressible, homogeneous;
- air movement does not influence the wave motion.

It is common practice to split the relative water depth d/L into the following zones:

- shallow water $\quad \dfrac{d}{L} \leqq 0.04$

- transition zone $0.04 < \dfrac{d}{L} < 0.5$ $\hspace{4cm}$ (9)

- deep water $\quad \dfrac{d}{L} \geqq 0.5$

The relationship between wave length L, wave period T and celerity C is as follows:

$$C = \frac{L}{T} \hspace{4cm} (10)$$

This formula is valid in general, independent of wave height or water depth.

The velocity with which a group of waves progresses is in general not equal to the celerity of the individual waves within the group. While the wave crests move at a celerity C, the whole group moves at a group velocity C_g. Both theory and experience indicate that in deep water the group moves at a velocity which is half that of the single wave celerity.

In very shallow water the single wave celerity and the group velocity will become equal. For practical application, the important formulae have been summarised in Table 1.

Table 1. Most important formulae of the linear wave theory.

	shallow water $\frac{d}{L} \leqq 0,04$	transition zone $0,04 < \frac{d}{L} < 0,5$	deep water $\frac{d}{L} \geqq 0,5$
wave celerity	$C = \frac{L}{T} = \sqrt{gd}$	$C = \frac{L}{T} = \frac{gT}{2\pi} \tanh \frac{2\pi d}{L}$	$C = C_0 = \frac{L}{T} = \frac{gT}{2\pi}$
group velocity	$C_g = C = \sqrt{gd}$	$C_g = \frac{1}{2}\left(1 + \frac{\frac{4\pi d}{L}}{\sinh \frac{4\pi d}{L}}\right)C$	$C_g = \frac{1}{2}C = \frac{gT}{4\pi}$
wave length	$L = T\sqrt{gd} = CT$	$L = \frac{gT^2}{2\pi} \tanh \frac{2\pi d}{L}$	$L = L_0 = \frac{gT^2}{2\pi} = C_0 T$

11.2.2 Local characteristics of the individual wave field

Irregular waves are much more difficult to describe than regular waves. The chaotic character of waves produced by wind is the real feature of those waves. This apparent chaos can only be put into some sort of order by expressing certain phenomena in terms of probabilities of occurrence. Hence it is necessary to make use of the probability theory.

The usual statistical parameters can be employed to describe wave distribution. In practice for coastal defence works, the smaller waves are often neglected, and the average of the highest 1/3-rd part of the waves is adopted as the significant wave height. This significant height is approximately equal to the wave height estimated by an experienced observer. Its value is commonly indicated by H_s.

A disadvantage of H_s is that it gives only a very approximate description of the total wave pattern. Because many (stochastic) processes governed by the rules of probability do not have an arbitrary nature, these can be described by means of theoretical distribution functions.

Wave heights of irregular wind waves can be described, with a reasonable degree of accuracy, by the Rayleigh distribution function (see Fig. 22):

$$P_r(\underline{H} > H) = e^{-2\left(\frac{H}{H_s}\right)^2} \tag{11}$$

where:

$P_r(\underline{H} > H)$ = the probability of exceedance of the wave height H
$\underline{H}$ = individual wave height (stochastic value) (m)
H = individual wave height (m)
H_s = significant wave height (m)

Wave period and length in an irregular wave system are rather more complicated than the height. For practical purposes, the period has been defined as the mean period $\bar{T}$, which is the mean value of the time T between two zero level crossings, or the significant wave period T_s pertaining to the highest 1/3-rd part of the waves. For the description of irregular waves use can thus be made of two parameters: firstly the significant wave height H_s and secondly the mean wave period $\bar{T}$ or the significant wave period T_s. Another possibility to describe the wave pattern is the wave spectrum; in contrast to the significant wave height, this fully presents the statistical properties of irregular waves. In this method, the energy density E is defined as a function of the frequency f of the spectral components.

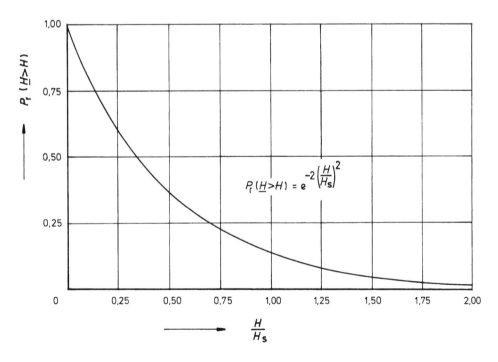

Fig. 22. Curve representing the Rayleigh distribution function.

11.3 Wave deformations

11.3.1 *Wave deformations in front of embankments*

Waves as they approach the (Dutch) coast undergo in general substantial deformations before they reach the toe of the dike. This is due to:

Refraction
When a wave enters water with a different depth, the celerity changes, as do the length and height so that under circumstances whereby the wave direction does not coincide with the direction of the change in depth, the direction of the waves also changes. One of the most important aspects on which refraction calculations can provide an answer is the question at which locations in front of the coast wave crests will meet and, through convergence, create increased wave motion in front of the dike.

Diffraction
Without changing wave length or period, the wave bends around coastal projections or dams. This phenomenon is usually attended and dominated by refraction.

Shoaling
Shoaling of waves denotes the process of transformation of the waves as a result of propagation in regions of varying water depth, where the wave direction runs parallel to the shoaling. Where the bed level gradually rises, the wave length reduces whilst the wave height increases; the wave period remains constant.

Bed friction
This can be neglected in relatively deep water, but near the coast the loss of energy can markedly reduce wave energy. However, in many situations the influence thereof is small in comparison with the phenomena referred to above.

Local wind
In contrast to energy losses through bed friction, local winds can add energy. A wave field approaching the coast can in this way absorb an appreciable amount of energy over, for example, the final 10 kilometres.

Currents
Wave deformations are also caused by changes in currents. Waves have a tendency to converge to locations where their celerity is reduced the most or to diverge to locations where their velocity is increased.

Wave breaking
A high-level foreshore in front of the dike will reduce the wave attack on the dike to a greater or lesser degree. The solitary wave theory expresses this for very shallow water

with a fixed ratio between the depth d_b where the waves break and the height of the breaking wave H_b:

$$H_b = 0.78 d_b \qquad (12)$$

The depth d_b denotes in this context the vertical distance from the bottom to the wave trough, which for this type of wave with the solitary wave characteristic nearly coincides with the still-water level at that location.

A formula much used for irregular waves is:

$$H_{s\ max} = 0.5 d_b \qquad (13)$$

where:

$H_{s\ max}$ = maximum significant wave height after breaking
d_b = vertical distance from the bottom to the still-water level

A high-level foreshore or sandbank in front of the dike therefore acts as a sort of filter for the waves: only the smaller waves pass through.

11.3.2 Wave breaking on the dike slope

When waves break on a dike slope, various breaker shapes can be distinguished, depending on wave steepness and the slope gradient. The breakers will also differ in energy dissipation and exerted forces.

A characteristic parameter which is used for various breaker types is the wave-breaking parameter:

$$\xi = \frac{\tan \alpha}{\sqrt{\dfrac{H}{L_0}}} \qquad (14)$$

where:

α = slope angie
H = wave height
L_0 = wave length in deep water $= \dfrac{gT^2}{2\pi}$
g = gravitational acceleration
T = wave period

The different types of breaker can be roughly classified according to the following table (smooth slope):

$\xi < 1$ spilling
$1 < \xi < 2.5$ plunging
$2.5 < \xi < 3.2$ plunging-collapsing   increasing wave steepness
$3.2 < \xi < 3.4$ collapsing-surging   decreasing slope gradient
$3.4 < \xi$ surging

The wave-breaking parameter ξ is also used to describe the stability of placed concrete block revetments.

Spilling breaker: $\xi < 1$
Breakers of the over-foaming type (spilling breaker) occur on very flat bottom gradients (see Fig. 23). These breakers continue running up over a substantial distance, losing energy continuously through breaking with foam formation at their crests, until these have disappeared altogether.

Fig. 23.   Shape of the spilling breaker.

Plunging breaker: $1 < \xi < 2.5$
Breakers of the overturning type (plunging breaker) occur on somewhat steeper bottom gradients and when the development of the wave crest in shallow water is not appreciably disturbed by other effects, such as wind, crossing waves, currents, irregularities on the bottom, etc. (see Fig. 24). The overturning breaker is characterised by the occurrence of a water curtain which is as it were "poured down" and detached from the front of the wave crest. Such breakers have, after turning over, which takes place in a short period of time, only little residual energy left.
On a comparatively steep bottom gradient the wave deformation process, until final breaking, takes place over a shorter distance. There is thus less chance of disturbing effects than on a flatter bottom gradient, and the steeper gradients have therefore an advantage in general for the development of overturning breakers.
Furthermore, a small initial wave steepness also encourages the development of overturning breakers, provided that other circumstances do not oppose such action. With a small initial wave steepness, the solitary wave development continues further and eventually attains a greater height, starting out from a given initial height, so that a greater "lift" of the crest occurs than can be reached from initially steeper waves.

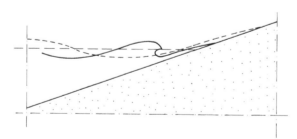

Fig. 24.   Shape of the plunging breaker.

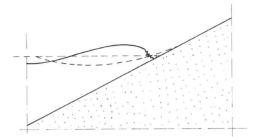

Fig. 25.   Shape of the collapsing breaker.

Collapsing breaker: $\xi \approx 3.2$

Breakers of the nearly overturning type (collapsing breaker) come between the plunging and the surging breaker types (see Fig. 25). In this form of breaker a vertical crest exists which has only partially collapsed.

Surging breaker: $\xi > 3.4$

Heaving breakers (or surging breakers) can be observed when the bottom or slope gradient is steeper still (and particularly when the wave steepness is very small) (see Fig. 26). The front of the wave crest is then, as it were, lifted up on the slope before the wave can turn over and a vertical up-and-down water movement is created with a comparatively thin foaming layer of water. The wave energy is now mostly reflected.

These surging breakers actually constitute the stage between the plunging breakers and the non-breaking waves whereby all wave energy is reflected.

If the energy delivery of the plunging breaker is compared qualitatively with that of the spilling breaker, the picture shown in Fig. 27 is obtained. It has to be realised here that in the foregoing diagrams the reality has been substantially simplified.

The considerations presented above are most accurately applied to regular waves which progress towards a flat and smooth slope; a situation which can be achieved in a laboratory. In the natural situation there are many factors which can disturb the picture presented above.

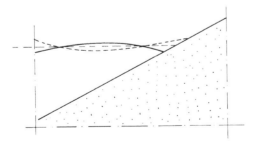

Fig. 26.   Shape of the surging breaker.

75

The wave impact caused by the plunging breaker occurs at a depth which varies between $\frac{1}{3}H$ and $\frac{2}{3}H$ below still-water level. The size of the wave impact is difficult to determine on a theoretical basis. The reasons for this are:

- When a wave breaks on a slope, a volume of air can be enclosed between the wave front and the slope. The thinner the air cushion the greater the maximum pressure and the shorter the duration of the impact.
- It is also possible that wave impact takes place into the return flow of the previous wave. The thicker the layer of the returning flow, the smaller the wave impact. The influence of the return flow increases as the slope angle decreases.

Fig. 28 shows a schematic diagram of a plunging breaker. If the revetment has some degree of permeability, the external wave load forces will be able to penetrate partially into the filter under the revetment, with the pressures underneath having a phase difference with the pressures on top of the revetment. The characteristics of pressure propagation in the filter are, however, also difficult to model or to describe theoretically.

Much research has been carried out into the wave impact phenomenon; nobody has as yet succeeded to describe it adequately, so that no theory can be indicated which can be applied properly. In order to obtain an impression of values, some measured data of the wave impact size are presented in Table 2. These values apply to a smooth impermeable slope. The duration of the pressure peak is in the order of 0.05 to 0.25 second.

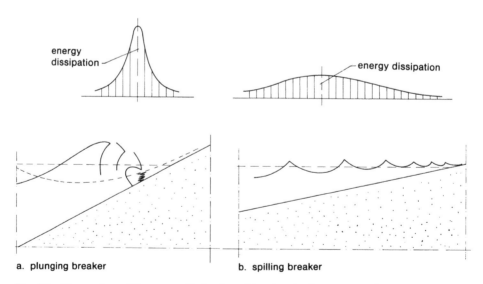

energy dissipation

energy dissipation

a. plunging breaker

b. spilling breaker

Fig. 27. Comparison of the energy dissipation of the plunging breaker and of the spilling breaker.

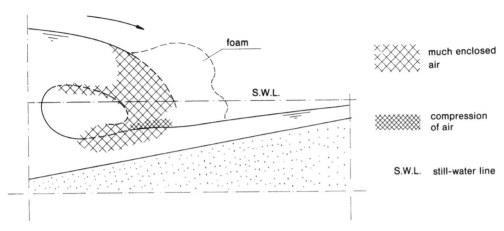

Fig. 28.   Diagram illustrating the descent of the breaker tongue.

For a description of the wave run-up phenomenon, reference is made to [22] and [25, part XV].

Table 2.   Measured values of the size of the pressure shock at a wave impact.

slope angle	pressure impact in metres of water head
1 : 2	$2.3H$
1 : 3	$2.7H$
1 : 4	$2.3H$
1 : 6	$2\ H$

CHAPTER 12

PARTICULAR LOADS

Amongst the particular loadings can be counted:

- wreckage and deck loads from ships;
- drifting ice;
- floating debris;
- recreation;
- vegetation and sea organisms;
- shipping accidents.

Wreckage and deck loads from ships

Depending on the dimensions of wreckage and deck loads, such objects when carried by waves can exert great forces on revetments. For normal practice it is, however, often not too bad; it is normally within acceptable limits so that any damage can be repaired as part of the normal maintenance.

Drifting ice

Drifting ice is something which occurs much more frequently in lakes and rivers than along the sea-coast, and for lakes and rivers can give rise to damage of embankments. In order to obtain a reliable basis for the considerations in predicting future ice conditions, an accurate analysis of existing data is required in relation to the factors which influence ice formation. The following aspects are important:

a. Factors which determine water movement:
   - hydrographic data: topography (surface and depth), exchange of water between the sea and the concerned estuary;
   - hydraulic data: water levels (tides), velocities and tidal capacities.
b. Factors which influence ice formation:
   - climatological factors: air and water temperature, wind, etc.;
   - water salinity;
   - hydrographic factors;
   - ice supply from the river.
c. Ice observations.

On a lake and in saline water in the shallower areas, an ice field of great thickness can develop, which can locally increase appreciably where layers drift on top of each other. In general, the ice fields start to move under influence of the wind after thaw sets in and this can cause serious damage to revetments.

In case of a lake, the situation is more serious because the point of attack of the ice layers on the dike will be at or just below water level, i.e. at the toe construction and the adjoining lower slope. In this context the strength of the ice is important also; saline water ice is usually weaker (softer) than fresh-water ice. Parts of the toe construction including the dumped stone lower slope protection could be picked up by the ice.

Such ice fields will not be stopped by dike slopes or obstacles on the slopes. The ice field will fit the water level contour along the slopes accurately and anything protruding above the slope will (probably) be sheared off.

In this connection it is pointed out that projections from a placed block revetment, such as rows of piles or raised blocks to curb wave run-up, form points of attack for the large forces which a moving ice field can exert. As a result of such forces, damage need not be limited to the projecting elements, but can extend to the slope protection by lifting out of elements and even to seriously affecting the structure of the revetment.

In order to limit ice damage as much as possible, it is advisable to finish slopes and their revetments as smooth as possible.

Floating debris

In general, floating debris will not cause any damage to hard revetments. Grass covers may, however, be choked where the debris covers it. On rivers damage can be caused by tree-trunks.

Recreation

Concrete revetments in general suffer little damage from this source, but grass covers can suffer from certain forms of recreation. Anglers can occasionally cause damage by permanently removing a single element.

Vegetation and sea organisms

Vegetation can usually obtain a toe-hold in the joints between concrete blocks. Such vegetation is in general not damaging to the quality of a concrete block revetment. In the zones which are regularly awash, the growth of algae and barnacles occurs; the higher-order growth occurs especially where areas are not regularly awash.

Shipping accidents

A storm has a direct influence on shipping disasters; chances of coincidence are real. This has to be taken into account along busy shipping routes.

In such accidents the contribution of the revetment to the stability of the sea defence against such loads is not large. When there is a demand that the sea wall must be able to resist such extreme loadings, the main resistance will have to be found in the mass of the sea wall body.

CHAPTER 13

STABILITY OF THE HAND-SET BLOCK REVETMENT

13.1 General

The stability of a revetment comprising placed (concrete) block elements is influenced by several variables, such as:

a. The characteristics of the construction (revetment and underlayer):
   - weight and/or dimensions of the block elements;
   - volumetric mass of the material;
   - friction between elements and the underlayer and between the elements themselves;
   - compressive forces (prestress) in the plane of the revetment;
   - interlock between the individual elements;
   - permeability of revetment and underlayer;
   - slope angle and shape of the slope;
   - sandtightness and erosion resistance of the underlayer;
   - influence of the transition constructions on the strength of the revetment;
   - roughness and water-storage capacity of the revetment;
   - long-term characteristics.

b. The hydraulic boundary conditions:
   - wave spectrum (wave heights and periods);
   - grouping of waves;
   - angle of wave attack;
   - breaker type and breaking location;
   - wave run-up;
   - wind effects;
   - wave deformation in front of the sea wall;
   - currents;
   - instantaneous water level;
   - frequency of occurrence of a specific hydraulic boundary condition on a specific location on the slope.

This summary of variables is probably not complete, but is does indicate the size of the problem. The quantitative effect of many of these variables is as yet insufficiently known.

The failure of a construction or part of a construction occurs if the loading exceeds the strength. As with traditional building constructions, the question may be direct failure due to excessive loading, and a failure due to moderate but frequently occurring loadings (e.g. erosion). The first type of failure is of importance to the revetment, and the second type to the underlayer and hence indirectly to the revetment.

The notion of stability has to be viewed for individual elements and for the slope as a whole. The resistance against lifting out a single element relies on its weight, possibly increased by frictional forces exerted by adjoining blocks. In case the frictional forces between elements are large, or if inter-connection is obtained in another manner (inter-lock systems), then the stability is determined not by the single element but by the slope protection as a whole. Where blocks have been placed alongside each other without interlock or joint-filling materials, a safe approximation is to calculate the stability on the basis of the loose block. The first considerations presented here will concern the loose-lying element, after which some thought will be given to the stability of interlock or friction revetments.

13.2 Loose-lying elements

It is necessary to differentiate between elements placed on a permeable or on an impermeable subsoil.

13.2.1 *Permeable underlayer*

For the disturbance of the stability of a loose element on a permeable base, present views distinguish about 8 possible mechanisms (see Fig. 29):

a. When wave run-up has reached its maximum level, the flow returns under influence of gravity. During this return flow, pressures on the revetment reduce. When the revetment is (hydraulically) rough, the return flow can result in current forces, inertia forces and lift forces.

b. Depending on the permeabilities and the geometry, the water which has penetrated through the revetment cannot flow back immediately, which results in forces attempting to lift the revetment. In general the height reached by the wave run-up is greater than the depth to which the wave withdraws below still-water level. Therefore, the water penetrates over a greater surface area into the filter than the surface area where outflow occurs. This results in a rise of the phreatic line in the filter and, consequently, an increase in the pressures underneath the hand-placed revetment. This effect is cumulative for a number of waves.

c. When the next wave arrives at the slope, the pressures on the slope increase under this wave. These pressures can be transmitted through the filter right in front of the wave, again producing uplift pressures under the revetment. Such pressures will only occur over a limited area adjoining the wave front.

d. Following the phase described under c, there are also substantial changes in the velocity field due to the approaching wave. The streamlines become curved, comparable with eddies. As a result of the curved velocity field the pressures above the revetment can be reduced.

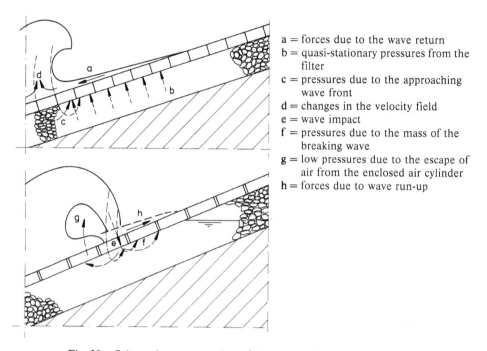

a = forces due to the wave return
b = quasi-stationary pressures from the filter
c = pressures due to the approaching wave front
d = changes in the velocity field
e = wave impact
f = pressures due to the mass of the breaking wave
g = low pressures due to the escape of air from the enclosed air cylinder
h = forces due to wave run-up

Fig. 29. Schematic representation of the possible failure mechanisms.

e. A breaking wave will produce wave impact on the slope, with sharply rising pressures with a duration of approximately 0.05 to 0.25 second. These pressures on the revetment can be transmitted through the filter and cause short-term pressures under the revetment.
f. After this short-lived phenomenon, a large mass of water falls on the slope. The high pressures can be transmitted under the revetment, just before the point where the wave breaks, and thus result in pressures which will lift the revetment.
g. After the wave makes contact with the slope, a large reduction occurs in the pressures on the revetment (even negative pressures, i.e. below atmospheric pressure). This phenomenon is explained as being caused through oscillations of the air "cylinder" enclosed in the breaking wave.
h. After the wave breaks, wave run-up takes place. During this phase, pressures on the revetment increase. At that stage, however, no critical circumstances exist; except when the revetment is rough or when blocks have been partly lifted out from the revetment. In these cases current forces, inertia forces and lift forces occur.

It can be concluded from research carried out at the Delft Hydraulics Laboratory and the Delft Geotechnics that, under the circumstances prevailing during the tests, the failure mechanisms b "quasi-stationary pressure differences" and c "pressures due to the approaching wave front" are very important to the stability of a block revetment. The mechanisms discussed can, however, not be viewed separately; a combination of

Row of units forced upwards by pressure from within the filter.

Pattern of damage associated with individual stability of the units.

Damage at the transition from a basalt to a concrete slope.

mechanisms can occur. For example, both the wave characteristics and the slope angle will have an influence on the importance of the various mechanisms.

Results of computer calculations for mechanisms b and c are shown in Fig. 30. For the calculation, the time was stopped at the moment the wave breaks, whilst the phreatic line is near the still-water level. To the left of the wave-breaking point, the excess pressure of the breaking wave pushes water through the revetment into the filter (mechanism c). On the right-hand side flow takes place from the elevated phreatic level (mechanism b). Both mechanisms influence each other. The revetment stability is affected more when both mechanisms b and c occur together than when they occur separately.

From the differential pressures (pressure above revetment minus pressure below revetment) shown in Fig. 30, it is clear that the revetment immediately to the right of the wave-breaking point is potentially unstable, which means that the differential pressure is greater than the weight of the revetment.

The mathematical approach illustrated here is, however, too cumbersome for practical purposes; for further information one should consult the CUR/COW report "Background to the guide to concrete dike revetments" [32].

The foregoing considerations make it clear that it is not a simple matter to produce a

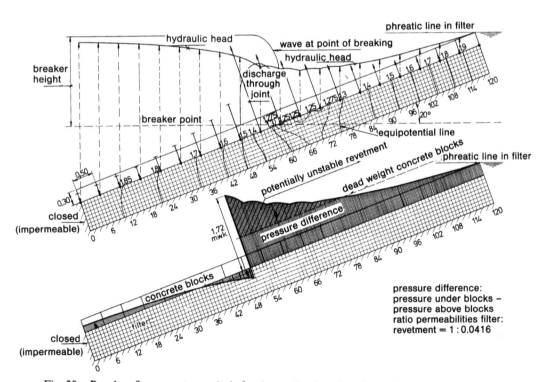

Fig. 30. Results of a computer analysis for the mechanisms b and c at the moment the wave breaks.

program for an analytical computer model, which can be used simply in practice, for all possible failure mechanisms; this has so far only been reasonably successful for mechanisms b and c, using the necessary schematical representations.

The analytical computer model has indicated that the permeability ratios between the revetment and the underlying filter are of great importance. This is expressed in the dispersion length λ:

$$\lambda = \sin \alpha \sqrt{\frac{kbd}{k'}}$$

where:

α = slope angle of the revetment
k = permeability of the filter layer
k' = permeability of the block revetment
b = thickness of the filter layer
d = thickness of the block revetment

The working-out of the computer model for mechanism b provides a relationship between the maximum difference in hydraulic head below and above the block work $\Delta\Phi$ and the vertical water level variation H. An indicative example of this is given in Fig. 31. It can be seen from this figure that a permeable revetment (k' is large) and a dense filter (k is small) have a beneficial influence on the stability ($\Delta\Phi$ is small). Also, a relatively thin filter layer (b is small) produces less high pressures under the blocks.

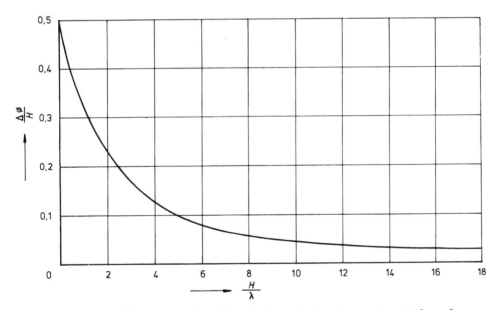

Fig. 31. Maximum difference in hydraulic head below and above the revetment $\Delta\Phi$ as a function of the dispersion length λ.

Similar conclusions are valid for the simultaneous occurrence of mechanisms b and c. These qualitative results correspond with the results of the tests carried out by the Delft Hydraulics Laboratory.

The various mathematical models which have been developed only offer application on a limited scale for the calculation of the weight of a revetment. For the time being the results obtained from laboratory tests will have to be used.

Through the application of the combined research data referred to in the literature, an attempt can be made to form an empirical formula. By using a much simplified equilibrium examination a formula can be derived. The coefficients employed will then have to be checked against the test results.

An example of this is given in Fig. 32, where the wave breaking parameter ξ (see also Section 11.3) is plotted against the parameter $H_s/\Delta d$. The equilibrium consideration at right angles to the slope gives the formula:

$$\frac{H_s}{\Delta d} = \frac{\cos \alpha}{c\xi}$$

where:

H_s = significant wave height

α = slope angle of the revetment

Δ = relative density of the block $\dfrac{\varrho_b - \varrho_w}{\varrho_w}$

ϱ_b = volumetric mass of the block

ϱ_w = volumetric mass of water

d = thickness of the block

ξ = wave-breaking parameter $= \dfrac{\tan \alpha}{\sqrt{\dfrac{H_s}{L_0}}}$

L_0 = wave length in deep water; based on the significant wave period

c = coefficient to be checked against test results

On the assumption that cos α is approximately equal to one (cos $\alpha \approx 1$), Fig. 32 represents a somewhat "safe" approach with $c=0.25$. The influence of the wave breaking parameter ξ on the stability of the placed blocks is clearly indicated.

Apart from the thickness d of the revetment, the relative density Δ of the blocks is also of great importance to the stability. Through a relatively limited increase of the volumetric mass ϱ_b, the under-water weight of the blocks increases markedly.

The "black-box" approximation presented here is fairly "rough"; thus, for example, the effect of the permeability ratio of revetment to filter is not accounted for, nor are the surface block dimensions and the thickness of the filter.

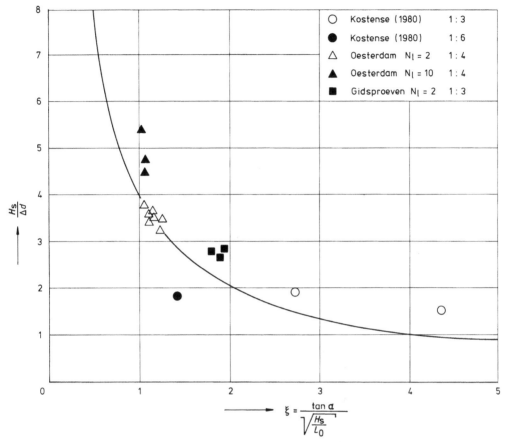

Fig. 32. Wave-breaking parameter ξ as a function of the parameter $H_s/\Delta d$ for irregular waves and a permeable foundation layer.

13.2.2 *Impermeable underlayer*

From tests carried out by the Delft Hydraulic Laboratory for the design of the Oester-dam, in the "Delta flume", it appeared that a revetment of blocks placed on a clay base layer possesses greater stability than is the case for block revetments on permeable base soils. The explanation is that the pressure build-up under the revetment is made more difficult by the impermeable character of the clay. As soon as a block is lifted slightly by an initial uplift pressure, water will have to flow under the block to fill the hollow space; this is also less easy with a clay base than with a granular filter underlayer.

However, in order to permanently benefit from the increased stability, the demand on the clay quality needs to be high (see Chapter 5). Especially the presence of sand-lenses is detrimental, whilst application in the tidal zone can be problematical.

Little is known so far about the failure mechanism.

87

13.3 Interlocked or friction block revetments (tightly fitting blocks)

In this case, equilibrium considerations concerning uplift should not be related to only one block, but to a number of blocks. Computations for the stability of the slope as a whole are made more difficult by:

- the size of the area which will more or less act as a "slab" is not known;
- lack of knowledge on the degree of pressure between the blocks (prestressing);
- lack of information on the deformation characteristics of the "slab";
- the indistinct nature of the failure mechanism.

Especially in case of placing revetments with a reliable interlock and hence high stability under wave attack, the strength of the other parts of the construction should not be neglected. It is, for example, possible that the waves acceptable to the block revetment would develop large pressure variations in the underlying filter. This could cause washing out of material, migration from the filter parallel to the slope and penetration of the sand from the body of the dike could occur, which in the long term could cause large settlement or undermining of the revetment.

13.4 Summary of the research in the "Delta flume"*

Various large-scale experiments (blocks to scale 1:2) were carried out in the Delta flume of the Delft Hydraulics Laboratory (at De Voorst). Figures 33 and 34 give a summary for regular waves and irregular waves respectively. Translating the results of the tests with regular waves to those with irregular waves is not yet possible.
The figures present the (dimensionless) wave height, which produces the damage, plotted as a function of the wave breaking parameter ξ (corresponding to Fig. 32). The revetment type and thickness, the wave period and the slope angle are given in the figures. From the figures it can be found that:

- Variation of the permeability of loose blocks on a filter had little effect on the stability. The revetment was, therefore, apparently sufficiently permeable in all cases investigated (see Fig. 34).
- Blocks on good clay have greater stability, which needs the additional comment that in short-duration tests the stability increased even more:
  $H_s/\Delta d=6.9$ (see Fig. 34).
- Blocks with a well-defined interlock possess a very large stability (see Figs. 33 and 34). The behaviour of the underlying filter can, however, be a limiting factor.

With the Armorflex and Basalton blocks the stability is appreciably increased by filling the spaces between the blocks with granular material, which increases frictional forces. It is not even possible to give the exact failure limit as the Delta flume is incapable of

---

* The Delta-flume, see plate page, is a very large wave flume built especially for research work connected with the Delta Project.

Model investigation into the stability of dike revetments of concrete blocks in the Delta flume of the Delft Hydraulics Laboratory at De Voorst. The photograph shows a breaker just about to hit the slope.

producing sufficient force to make the model block slope fail for this type of revetment. This research also showed that, when the spaces between the blocks are filled with granular material, the exact form or shape of the blocks is not of great importance. For all the block shapes investigated, the failure limit proved to be beyond the wave flume capacity, due to the filling of spaces with granular material.

Although the stability of the blocks in the figures has been given as a function of two

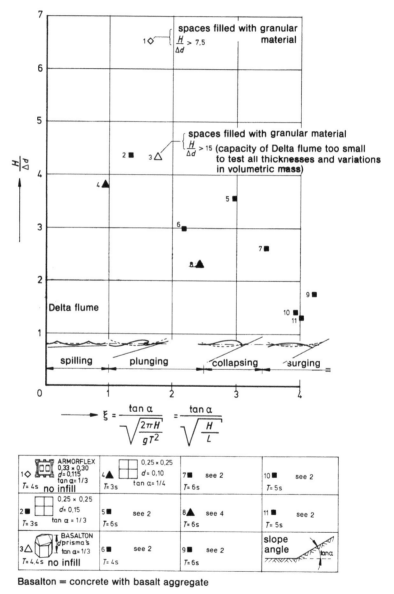

Fig. 33. Stability values of various block revetments on a filter base for regular waves.

90

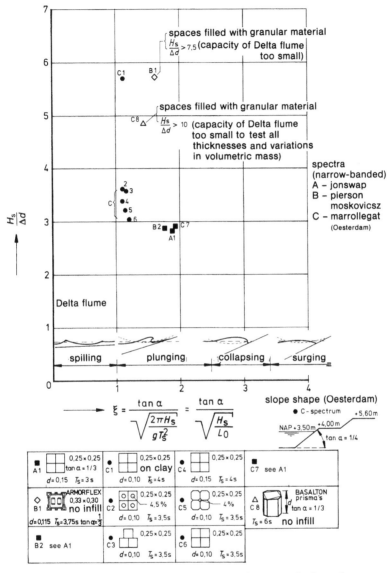

Fig. 34. Stability values of various block revetments on a filter base for irregular waves (except test C1, blocks placed on clay).

dimensionless parameters, it is nevertheless not permissible to extrapolate the results shown to areas which are clearly beyond the measured values presented in the figures, such as for example a different slope angle. The reason is that the parameters $H/\Delta d$ and ξ only very partially represent the failure mechanism. Some caution in the application of the results of the Delta flume tests is therefore essential.

91

For Dutch estuaries (without taking swell into account), the wave steepness H_s/L_0 is often found to vary between 0.035 and 0.065. When an average value of 0.05 is selected for H_s/L_0, then the ξ values are sharply limited:

$$\tan \alpha = \tfrac{1}{3} \rightarrow \xi = 1.5$$
$$\tan \alpha = \tfrac{1}{4} \rightarrow \xi = 1.1$$
$$\tan \alpha = \tfrac{1}{5} \rightarrow \xi = 0.9$$
$$\tan \alpha = \tfrac{1}{6} \rightarrow \xi = 0.75$$

For these ξ values Fig. 34 shows no lower value than 3 for $H/\Delta d$ in case of loose blocks on a filter layer. This may therefore be taken as a lower limit.

For loose blocks on clay a minimum value of 5 can be adopted for $H/\Delta d$, provided that the clay criteria described in Chapter 5 have been satisfied.

With systems having a good and trustworthy interlock, higher values have been found. The performance of the underlayer can, however, be a limiting factor.

CHAPTER 14

SAFETY CONSIDERATIONS

14.1 General

Loads and strength parameters can in general not be predicted in advance as to the precise values they will have in a structure. The loads and the strength of a structure in practice are evidently subject to variations and will, therefore, have to be taken as random variables.

A random variable is defined as a variable which is not characterised by a single value, but is described in terms of statistical parameters.

The opposite of a random variable is the deterministic variable, of which the value is known with certainty. In practice deterministic variables for loads and strengths do not occur.

Sometimes the spread of values which occurs is, however, so small that it is justifiable to consider the variable concerned as a deterministic one. Like the loads and strengths, the dimensions of structures are also stochastic.

However, practice shows that most of the failure cases on record are not caused by stochastic variations in strength and/or loads, but by constructional deficiencies and human errors. This concerns, for instance, the possibility of calculation errors in the strength computation or, worse, not recognising a normative failure mechanism.

Many failures occur during construction, whilst errors during execution in general form a source for possible construction failures. Neglect of the necessary inspection and maintenance can also have serious consequences.

The following discussion is only intended to highlight the philosophy and possibilities of some fairly recently developed methods. Finally, a few remarks will be made about the levels of safety to be adopted.

14.2 Description of probabilistic methods

A reliability function Z is defined as the difference between the strength or resistance of a structure and the load:

$$Z = R - S \tag{15}$$

where:

Z = the reliability function
R = the strength of the structure (resistance)
S = the load on the structure (solicitation)

In order to obtain a structure which serves its purpose, the value of Z should be greater than zero; no failure would occur in that case.

Strength R and load S will in their turn in general be functions of other variables. The reliability function can be written in the form:

$$Z = R(X_1, ..., X_k) - S(X_{k+1}, ..., X_n)$$

It is customary to denote the variables $X_1, ..., X_n$ in the reliability function as "base variables".

It should be noted that the reliability function Z need not only be viewed as a formula; Z may also come out of a complex mathematical model, e.g. from a computer program. With dike revetments it is not known in advance what the normative situation will be, i.e. whether it will be water level, wave height, wave steepness, etc. If it is viewed from the probability of failure of the revetment, then often not the waves occurring during the most extreme water levels (super storm) determine the design criteria, but the waves at a lower level. This is because, although the wave attack at lower levels is less heavy, the probability of occurrence of a lower level storm is greater than for an extreme level storm. This aspect will only be shown in its true context with the aid of a probabilistic calculation, in which load and strength are treated as inseparable from each other.

The Delta Commission [29] in their design approach for sea and estuary embankments also introduced a stochastic element, but they returned, on practical considerations, to limiting this to indicating only the probability of exceedance of the loading. Primary sea-defences have to be designed, according to the Delta Commission, in such a way that they can *fully* withstand a storm surge with a given (selected) probability of exceedance. In practice the sea wall will often be designed in such a way that a balance exists between average strength and design load, leaving aside hidden safety factors. This means, however, that there is a 50 % probability that the embankment fails during the "design" storm surge, which is not in accordance with the requirements of the Delta Commission.

Fig. 35 shows schematically the method employed in the current (Dutch) Regulations for concrete, in which the probability distribution of the strength is also taken into consideration. With the aid of a central safety coefficient, sufficient distance is maintained between the "probability densities" of load S and strength R.

The core of the safety consideration is formed by the so-called probabilistic calculation. Generally speaking it means that the probabilities of failure are determined on the basis of the uncertainties in the imposed loads and the strength of the structure.

A probabilistic calculation in which these uncertainties are included in a purely formal manner, will soon lead to complicated or unsolvable mathematical formulations for the probability of failure. Various simplifications have, therefore, been gradually introduced. These simplifications refer to the way in which the uncertainties are incorporated.

To arrange the various possible procedures in a certain order, four levels of calculation have been distinguished, which vary from completely deterministic to completely probabilistic. These levels are:

94

Level 0:   A deterministic calculation. Both load and strength are given defined fixed values and the mathematical model is regarded as fact. By means of *one* overall safety coefficient all uncertainties are accounted for.

Level I: A semi-probabilistic calculation. The loads and strengths are based on characteristic values. By means of partial safety coefficients, i.e. coefficients which refer to separate variables, the remainder of the uncertainties are accounted for.

Level II: A probabilistic calculation, in which well-defined simplifications have been introduced in processing the stochastic variables; several methods are available for this.

Level III: A completely probabilistic calculation. The calculation is totally based on the theory of stochastics.

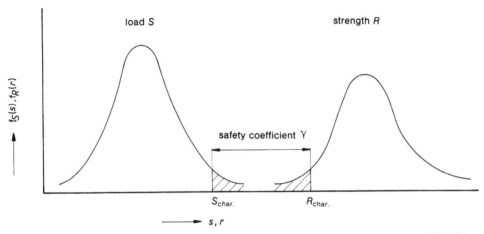

Fig. 35.   Design of a structure in accordance with the Regulations for concrete VB 1974/1984 (level I).

Essentially, the calculations on levels 0 and I are not probabilistic because the result of the calculations does not give a probability of failure. The values of the safety coefficients can, however, for the standard problems be derived from a calculation on level II or III. A calculation on level 0 and I can, therefore, implicitly refer to a particular probability of failure.

The considerations in this chapter will, because of the practical possibilities, be limited to probabilistic calculations on level II.

The simplifications which are introduced in level II calculations are primarily aimed at reducing a complex reliability function to a linear function. Subsequently, the distribution of the reliability function is approximated by replacing it with a normal distribution, of which the mean and the standard deviation are derived from the corresponding parameters of the base variables. The introduction of the simplifications in the reliability function can be done in various ways; two main methods are employed in practice:

– the mean value approach;
– the advanced method.

Mean value approach

In this method, the reliability function Z, for the mean of the various base variables, is developed in a mathematical series which is terminated after the linear terms.

Linearisation in the mean signifies in general that it is linearised in a point which does not lie on the failure limit $Z = 0$. This method is also not insusceptible to the manner in which the reliability function is formulated. However, on the other hand it can be said that the mean value calculation is relatively simple and can in many cases even be carried out completely by hand.

More complex iterative calculations are required to find a better design point, e.g. the following advanced method.

Advanced method

In order to overcome the objections against the mean value approach, improved reliability analyses have been worked out on Level II. The improvement in the advanced method concerns the choice of the design point.

Linearisation no longer occurs in the "mean" but instead in a point on the failure boundary. On the failure boundary, the design point is in addition chosen in such a way that the probability of occurrence of that value of Z is as large as possible. Where the base variables deviate from the normal distribution, care is taken to ensure that the replacing normal distribution has the same probability density and probability of exceedance. The definitive design point is determined by means of an iterative method. One consequence of the iterative process is that the calculation can in general no longer be carried out by hand.

14.3 **Load and strength**

It is necessary to know, for the formulation of the reliability function Z, the probability distribution functions and mutual relationships of the base variables. In this context the following points arise:

a. the probability distribution of the high water level;
b. the relation between wave height and high water level;
c. the probability distribution of the wave steepness;
d. the wave-breaking criterion for wave breaking on the foreshore;
e. the model for describing the stability of a placed revetment;
f. the probability distribution of the various parameters which determine the strength of the revetment, for example the slope angle, the block thickness and the clamping-friction between the blocks.

In the CUR/COW report "Background to the guide to concrete dike revetments" [32], the lead was given for a mathematical model in which the above points a to f have been incorporated. With the aid of a number of worked examples, the possibilities for a probabilistic approach have been illustrated therein.

In a probabilistic calculation the probability of failure is calculated as well as the design point. The design point comprises the optimum combination of base variables to make the probability that Z will be less than zero as large as possible. Thus, in addition to the probability of failure, information is obtained on the most dangerous water level in terms of probabilities of failure. This means that the revetment is safer for other water levels.

Information is also obtained on the contribution of every base variable to the variation of the reliability function Z. In this way it is possible to estimate to what extent the exact values of the probability distribution of the variable concerned are determining the final result. The introduction of a wave-breaking criterion enables the quantification of the favourable influence the breaking of waves on a high foreshore has on the stability. Fig. 36 illustrates some results of computer calculations using the advanced method derived from the aforementioned report. The ordinate of the graph presents the probability of failure and the abscissa indicates the water level at which the probability of failure is greatest. The lines in the graph were derived by varying the thickness of the revetment and the depth of the foreshore.

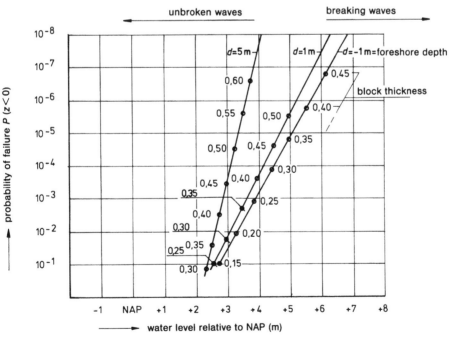

Fig. 36. Example of the results of a probabilistic calculation.

The figure shows the influence of the breaking wave, on a high-level foreshore, on the location of damage and the safety against failure. The boundary between breaking waves and non-breaking waves on a foreshore is influenced by the water level and the wave height. A link exists between water level and wave height in the sense that a greater wave height is related to higher water levels. These considerations lead to the conclusion (as shown in Fig. 36) that one can only indicate areas where waves break and areas where waves do not break; a distinct transition can not be indicated.

14.4 Safety level

In order to include the revetment weight in the probabilistic calculation method, the acceptable probability of failure has to be established. Between the daily practice in the sea wall construction industry and the design philosophy laid down in the Delta Report, there is sometimes a discrepancy with regard to the revetments. In practice, designs are sometimes made on the basis of experience, i.e. under actual usage conditions. The Delta Report in contrast, starts out from a design storm which lies outside the area of experience. There is a fairly wide gap between the two methods. In practice, occasional cases of damage are found at lower levels, say a probability of failure in the order of 10^{-2}. If this relatively large probability of failure were valid for the entire slope, it would mean that the revetment needs to be carried through on the slope to a much lower level than is usual at present, because the associated probability of a water level occurring at higher levels is smaller than 10^{-2}.

The Delta Commission has, however, given a design water level with a probability of exceedance of 10^{-4} for the calculation of the crest level. The sea wall still has to serve its purpose up to that level.

Between the uplift of a block at this high level and the failure of the sea wall, there still remains the mechanism of progressive erosion of the intermediate layers and underlayers and the body of the sea wall itself. It is not known how much additional safety this yields. The starting-point, however, has to be that, at a water level with a probability of exceedance of 10^{-4}, the revetment still has some residual strength. The accepted total probability of failure at the super storm level, according to the Delta Report, will then be in the order of less than 10^{-4}.

By way of illustration, Fig. 37 shows the course of the water levels of two storms, I and II; phase differences between tide and wind set-up have not been taken into account. Storm II lasts longer, reaches a higher water level and produces at its maximum greater waves than storm I.

It is assumed that up to peak I the two storms are identical. Storm I has, of course, a greater probability of occurrence than storm II. Because storm I diminishes in strength after peak I, and the sea wall at this (comparatively) low water level has, because of its great thickness, a large residual safety margin, it is perhaps acceptabel to have a relatively large probability of failure of the revetment. After the storm, the damaged part of the revetment and the dike body can be repaired.

98

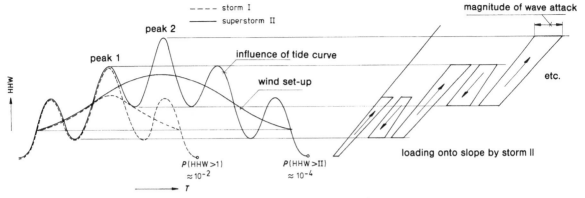

Fig. 37. Water levels and wave attack.

On this line of argument, higher probabilities of failure would be acceptable for the lower levels. This argument is, however, incorrect.

When storm II occurs there is no time available to repair the damage caused by peak I. The result is that this storm would, so to speak, "roll-up" the revetment from the bottom upwards, which would create the danger that through rapid erosion the revetment higher up on the slope would be undermined. This is not acceptable.

The damage occurring at the lower levels will result in damage higher up on the slope, despite the constructional safety of the higher level revetment. The strength of the chain lies in its weakest link.

It has to be concluded, therefore, that at all levels the same probability of failure in the order of less than 10^{-4} must be adopted. In consequence, in the Netherlands there should be rarely if ever any serious damage observed to revetment and underlayer.

CHAPTER 15

SUMMARY

This guide is intended for engineers and technicians directly associated with the design and management of dikes. It is not intended as a scientific work dealing exhaustively with theoretical fundamentals. It has been endeavoured as much as possible to give background information without offering a solution for every conceivable problem. For a more in-depth treatment of these matters, the reader is referred to the CUR/COW report "Background to the guide to concrete dike revetments" [32]. With a view to the application in the Netherlands, the treatment of the subject is confined to the type of revetment composed of relatively small units.

In the first part of this century, in-situ casting of concrete – then still a new material – was frequently carried out, not always with favourable results. In the course of time, the use of precast concrete units was increasingly adopted (Chapter 1). At first, manual methods were used for producing them and placing them on the slopes to be protected. The units were made in a wide variety of shapes. Mechanization was subsequently introduced into the production process, and the handling and installation of the units were also to a great extent mechanized. In conjunction with this development the complexity and variety in the shapes of the units were reduced.

For the revetment, i.e. the protective covering, of a flood defence requirements are formulated with reference to the purpose of the revetment, the technical features of constructing it, and possible special circumstances involved (Chapter 2). Various types of revetment are distinguished with reference to the properties of the units and of the underlayer (Chapter 3). The material-technological properties of concrete in a marine environment are only briefly considered, because comprehensive and readily accessible literature and codes of practice are available on the subject (Chapter 4).

Requirements are applied to the underlayers of the revetment because these are important in maintaining its stability under wave attack and in ensuring that the structure will fully and permanently serve its purpose (Chapter 5). In this connection, a distinction is drawn between permeable and impermeable underlayers. Research carried out at the Delft Hydraulics Laboratory has shown that an impermeable layer (clay) gives the revetment greater stability under wave attack than a permeable layer does. In order to derive lasting advantage from this greater stability, it is necessary to lay down requirements as to the material properties and the manner of use of the clay, while the circumstances of the job may impose restrictions on applicability.

It is indicated what materials can suitably be used for a permeable underlayer and what requirements they must satisfy, more particularly with regard to the penetration of material from the subsoil into the filter material.

100

Wave attack, which is a major factor governing the stability of the revetment, has a different frequency of occurrence at each level of the slope and varies in magnitude. This depends on the type of dike concerned, e.g. a sea wall or a lake embankment, and on many other factors. These matters are explained with reference to defined zones of loading (Chapter 6).

The shape of the cross-sectional profile of the dike is of influence on the type of element suitable for revetment construction (Chapter 7).

In the experience of many dike managers, substantial damage is liable to occur at the transition from one type of revetment to another and in zones where the revetment ends. Although it is not practicable to give standard solutions, outright mistakes can be highlighted. The toe construction, the upper boundary of the hard revetment and the transition to a different type of revetment are dealt with (Chapter 8).

The next two chapters are concerned with construction (Chapter 9) and with management and maintenance (Chapter 10).

The loads acting on the structure and its strength are then considered. First, the hydraulic boundary conditions such as wave characteristics, wave fields and wave deformations (including the breaking of waves) are dealth with (Chapter 11); next, particular loads are reviewed (Chapter 12).

As an interim result of long-term research still in progress, some information concerning the stability of the placed revetment is given (Chapter 13). On the one hand, results of theoretical model studies and, on the other, results of recent research in a wave flume (1 : 2 linear scale) are reported. Because of the complexity of the subject matter there is as yet no easy-to-use mathematical model available for dealing with various kinds of revetment and subsoil. All the same, with the aid of the data yielded by empirical research it is possible to approximately determine the thickness of one of the given types of revetment.

The magnitude and the location of the load acting on the revetment, as well as the strength of the dike revetment, are subject to variation. Particular values have a particular probability of being exceeded or of not being attained. Safety considerations are presented with reference to this (Chapter 14). It is indicated what approach might be possible with a view to obtaining a better understanding of the actual safety of a concrete dike revetment. Linking up with the work of the Delta Commission, an order of magnitude for the safety level is given.

REFERENCES

General

1. KLEY, J. VAN DER and H. J. ZUIDWEG, Polders en dijken (Polders and dikes). Agon Elsevier, Amsterdam/Brussel 1969.
2. AGEMA, J. F., Collegedictaat Waterkeringen, f3/f11, Afdeling der Civiele Techniek, Technische Hogeschool Delft (Notes on the lectures in water defences, f3/f11, Faculty of civil engineering, Delft University of Technology), 1982.
3. Die Küste. Archiv für Forschung und Technik an der Nord- und Ostsee, Heft 36, 1981. Herausgeber: Kuratorium für Forschung im Küsteningenieurswesen.
4. BEKKER, M. E., J. DE BOER and J. DE JONG, Kust- en oeverwerken (Coast and embankment works). Stam Technische Boeken, 1974.

History

5. COLIJN, P. J., Zee- en rivierwerken (Sea and river works). N.V. Uitgeversmaatschappij, voorheen Van Mantgem & De Does, 1921.
6. MURALT, R. R. L. DE, Dijk- en oeverwerken van gewapend beton volgens het "Systeem de Muralt" (Dike and embankment works in reinforced concrete based on the "de Muralt-system"). Technische Boekhandel en Drukkerij J. Waltman Jr., Delft 1913.

Concrete technology

7. NEN 3880, Voorschriften Beton VB 1974/1984 (Regulations for concrete). Nederlands Normalisatie-instituut, Delft 1984.
8. NEN 7024, Glooiingselementen van beton (Concrete slope-elements). Nederlands Normalisatie-instituut, Delft 1972.
9. CUR-rapport 22, Weerbestendigheid van beton (Durability of reinforced concrete). CUR, Gouda 1961[1].
10. CUR-rapport 64, Vorstbestandheid beton (Frost resistance of concrete). CUR, Gouda 1974[1].
11. CUR-rapport 90, Reparaties van betonconstructies (Repairs to concrete structures), deel I: Vervangen of repareren van beschadigde constructies. CUR, Gouda 1977[1].
12. CUR-rapport 96, Beton en afvalwater (Concrete and sewage). CUR, Gouda 1979[1].
13. CUR-rapport 99, Erosion van beton (Erosion of concrete). CUR, Gouda 1980[1].
14. CUR-rapport 100, Duurzaamheid maritieme constructies (Durability of maritime structures). CUR, Gouda 1981[1].

Characteristics of materials regarding underlayers

15. Kust- en oeverwerken, in praktijk en theorie (Coast and embankment works, practice and theory). Nederlandse Vereniging Kust- en Oeverwerken, 1979.
16. Kunststoffen en oeverbescherming (Synthetic materials and embankment protection). Uitgegeven onder auspiciën van de Nederlandse Vereniging Kust- en Oeverwerken, Stam Technische Boeken, 1975.
17. Kunststoffilters in kust- en oeverwerken. (Geotextile filters in coast and embankment works). Nederlandse Vereniging Kust- en Oeverwerken, 1983.
18. GRAAUW, A. F. F. DE and M. A. KOENDERS, Stand van zaken bij het onderzoek naar granulaire filters (State of the art of the research into granular filters). Waterloopkundig Laboratorium en Laboratorium voor Grondmechanica, Nota S 469, Delft 1980.
19. LAAN, G. J., De toepasbaarheid van mijnsteen in de waterbouw (Applicability of colliery shale in hydraulic engineering). WKE-R-78156, Rijkswaterstaat Deltadienst, werkgroep keuring bouwstoffen voor de waterbouw, 1980.
20. Interimrapport "Klei onder steenzettingen voor Oesterdam en Philipsdam" (Clay under placed block revetments; Oesterdam and Philipsdam). Rijkswaterstaat Deltadienst, werkgroep klei, 1984.

Hydraulic boundary conditions

21. Shore protection manual. Volumes I, II and III, U.S. Army Coastal Engineering Research Center, 1977.
22. BATTJES, J. A., Wave run-up and overtopping. Technical Advisory committee on water defences, report 1E, 1972.
23. Over het berekenen van deltaprofielen voor dijken langs de Westerschelde (Calculations of delta profiles of dikes along the Western Scheldt). Rijkswaterstaat, Directie Zeeland, Studiedienst Vlissingen, 1972.
24. GROEN, P. and R. DORRESTEIN, Zeegolven (Sea waves). Staatsuitgeverij, Den Haag 1976.

Stability hand-set block revetments

25. Taludbekleding van gezette steen (Slope revetment of placed blocks). Part I to XV, M 1975/ M 1881, Waterloopkundig Laboratorium and Laboratorium voor Grondmechanica, Delft 1982/1984.
26. Stabiliteit Armorflex-steenzetting onder golfaanval (Stability of Armorflex-block revetment under wave attack). Verslag modelonderzoek, M 1910, Waterloopkundig Laboratorium and Laboratorium voor Grondmechanica, Delft 1983.
27. Basalton, Stabiliteit onder golfaanval (Stability under wave-attack). Verslag modelonderzoek, M 1900, Waterloopkundig Laboratorium and Laboratorium voor Grondmechanica, Delft 1983.
28. Gobi-blokken als taludbekleding (Gobi-blocks as slope revetment). M 1184, Waterloopkundig Laboratorium, Delft 1973.

Safety considerations

29. Rapporten Deltacommissie, Deel 1 t/m 6 (Reports of the Delta Commission, Parts 1 to 6). Staatsuitgeverij, Den Haag 1960/1961.
30. VROUWENVELDER, A. C. W. M. and J. K. VRIJLING, Collegedictaat Probabilistisch ontwerpen, b3, Afdeling der Civiele Techniek, Technische Hogeschool Delft (Notes on the lectures in probabilistic design, b3, Faculty of civil engineering, Delft University of Technology), 1983.
31. CUR-rapport 109, Veiligheid van bouwconstructies. Een probabilistische benadering (Safety of structures. A probabilistic approach). CUR, Gouda, 1982[1].

Other

32. CUR/COW-rapport, Achtergronden bij de leidraad cementbetonnen dijkbekledingen (Background to the guide to concrete dike revetments). Centrum voor Onderzoek Waterkeringen, Den Haag 1984.
33. Leidraad voor de toepassing van asfalt in de waterbouw (The use of asphalt in hydraulic engineering. Technical advisory committee on water defences, Rijkswaterstaat communications No. 37, The Hague 1985). Technische Adviescommissie voor de Waterkeringen, Staatsuitgeverij, Den Haag 1984.

Further information can be found in:

34. BEZUIJEN, A., M. KLEIN-BRETELER and K. J. BAKKER, Design criteria for placed block revetments and granular filters. 2nd international conference on coastal engineering in developing countries, Bejing, 1987.
35. BEZUIJEN, A., M. KLEIN-BRETELER and K. W. PILARCZYK, Large-scale tests on block revetment placed on sand with a geotextile. 3rd international conference on geotextiles, pp. 501–505, 1986.
36. BURGER, A. M., M. KLEIN-BRETELER, L. BANACH and A. BEZUIJEN, Analytical design method for block revetments. 21th international conference on coastal engineering, Malaga, 1988.
37. PILARCZYK, K. W., Dutch guidelines of dike protection. 2nd international conference on coastal engineering in developing countries, Bejing, 1987.
38. PILARCZYK, K. W. and A. M. HENDRIKSMA, Concrete dike revetments. Betonteil + Fertigteiltechnik 4, 1987, pp. 262–269.
39. The closure of tidal basins, 2nd ed., Delft University Press, 1987.

---

[1] With a summary in English.

Milton Keynes UK
Ingram Content Group UK Ltd.
UKHW020820141024
449569UK00008B/495